THE DEATH
OF
EVOLUTION

1/14/03

"*Thine are the heavens, and thine is the earth: the world and the fulness thereof thou hast founded: the north and the sea thou hast created.*"

—Psalm 88:12-13

THE DEATH
OF
EVOLUTION

By

Wallace Johnson

"For there shall be a time, when they will not endure sound doctrine; but, according to their own desires, they will heap to themselves teachers, having itching ears: And will indeed turn away their hearing from the truth, but will be turned unto fables."

—2 Timothy 4:3-4

TAN BOOKS AND PUBLISHERS, INC.
Rockford, Illinois 61105

Nihil Obstat: J. A. Clarke, D.D., D.C.L.
                   Censor Deputatus

Imprimatur: ✠ Francis Rush
                   Archbishop of Brisbane
                   November 23, 1981

Originally published as *The Crumbling Theory of Evolution* by J. W. G. Johnson (Wallace Johnson) in Australia, 1982. Third printing (updated) in 1987. Published by Perpetual Eucharistic Adoration, Inc., Los Angeles, c. 1986 under the title *Evolution?* and distributed by TAN. Polish edition (translated by Prof. Maciej Giertych) under the title *Na bezdrozach teorii ewolucji* ("The Crumbling Theory of Evolution") published in 1989 by Wydawnictwo Michalineum, Warszawa-Struga. Retypeset and published in 2000 by TAN Books and Publishers, Inc., with permission of Perpetual Eucharistic Adoration, under the title *The Death of Evolution*.

ISBN 0-89555-664-2

Library of Congress Control No.: 00-131563

Printed and bound in the United States of America.

TAN BOOKS AND PUBLISHERS, INC.
P.O. Box 424
Rockford, Illinois 61105

2000

"By the word of the Lord the heavens were established; and all the power of them by the spirit of his mouth: gathering together the waters of the sea, as in a vessel; laying up the depths in storehouses. Let all the earth fear the Lord, and let all the inhabitants of the world be in awe of him. For he spoke and they were made: he commanded and they were created." —*Psalm* 32:6-9

CONTENTS

Bible rather than evolution—Civilization did not evolve—Culture degrades with remoteness—Feral children—Human language.

FOREWORD

I understand sincere evolutionists; I was one. I do not understand those who not only will not listen to counter-argument, but would prevent others from listening.

Ten years ago, to humor a friend, I read Father O'Connell's *Science of Today and The Problems of Genesis.* My belief in evolution disintegrated, accompanied by anger that a whole counter-argument had been kept from me for so long. At the same time I began to discern evolution's potential threat to religion. I resolved to equip myself to help others to hear the counter-argument which was being so effectively suppressed.

I studied and lectured and learned. The lecture, "The Case Against Evolution," grew to a two-hour session, and it was recorded. The recording was converted into a booklet in 1976 by Miss Paula Haigh, who was conducting the Catholic Center For Creation Research in America. To my extreme surprise, the modest printing was well received. It was reprinted in Australia in 1979. Out of that booklet has come this small book. It involved a complete rewriting of the original, and enlarging it with so much additional material that the result is a different book.

With gratitude I acknowledge the personal effort of Miss Haigh, without which the original text would not have seen print. Also, I acknowledge the collaboration of Mr. A.W. Mehlert of Brisbane, who has authored articles in the *Creation Research Society Quarterly* and the *Bible Science Newsletter.* His specialized research, especially in the increasingly confused field of "ape-men," was most helpful. Acknowledged, too, is the help received from the anonymous ones: illustrators, typists, and those who helped bridge the financial gap for publishing costs.

There is now a massive literature by anti-evolution scientists, and it is hoped that this book will provide a digest of their overall case in an easily readable form for lay people—and for experts, too, if interested.

Wallace Johnson
December 1, 1981

THE DEATH
OF
EVOLUTION

"Thou in the beginning, O Lord, didst found the earth: and the works of thy hands are the heavens. They shall perish, but thou shalt continue: and they shall all grow old as a garment. And as a vesture shalt thou change them, and they shall be changed: but thou art the selfsame, and thy years shall not fail."
—Hebrews 1:10-12

Chapter 1

WHY WE MUST FIGHT EVOLUTION

Attack and Counter-Attack

"Every attack on the Christian faith made today has, as its basis, the doctrine of evolution." (Newman Watts, author of *Britain Without God*).

More than a century ago, in England, an Anglican bishop's wife said: "I do hope that what Mr. Darwin is saying is not true; and if it is, I hope it does not become generally known."

Today the wheel has turned. Instead of the bishop's wife, it is now Darwinists who are worried, because creation scientists have shown that evolution is false. The Darwinists, in their turn, are hoping that this does not become generally known. The evolution-biased mass media is ensuring that it does not become known.

The dominance of evolution ideas deadened belief in Divine Creation and supernatural religion. So was born the phenomenon of the 20th-century, the *secular man*. Well-educated; inured to evolution; often a very decent person; but he never thinks of God.

An atheist, Renan, predicted that the collapse of the supernatural would lead to the collapse of moral convictions. Evolution's naturalism has ousted supernaturalism, and we can see moral convictions collapsing. The Christian culture is crumbling; and the *"Post-Christian era"* has begun. That is the final fruit of evolutionism.

In 1859, Professor Sedgwick of Cambridge warned Darwin that, through his evolution ideas, "Humanity would suffer a damage that might brutalize it and sink the human race into a lower state of degradation than any into which it has fallen since its written records tell us of its history."

> In time, the theory of evolution permeated human thought in almost every direction . . . The ultimate result was exactly what Sedgwick had said it would be, brutalization. The new doctrine very soon began to undermine religion. (Robert E. D. Clark, M.A., Ph.D. in *Darwin: Before and After*, 1948.)

"All this has in a great measure lead to agnostic and atheistic beliefs of the present day. Perhaps the worst of all is that the minds of the young have been singed by doubt." (Father D. Murray in *Species Revalued*, 1935.)

> Actually, the work of the evolutionists will be largely responsible for the perilous times which are ahead, for evolution has been a large factor in bringing about the widespread godless philosophy which is characteristic of our time, and which will become worse. (Ex-evolutionist, Bolton David-heiser, Ph.D. Zoology & Genetics, in *Evolution and Christian Faith*, 1969).

Almost unchallenged for a century, evolution completely changed world thinking and caused havoc in religion. Yet, the counter-attack which has begun is not by religious writers, but by scientists.

- *The Evolution Protest Movement* was founded by a few eminent scientists in England in 1932. It has produced a continuity of sound literature against evolution.
- *The Creation Research Society* began in 1963 in the U.S.A. with 10 scientists. It has grown rapidly to over 650 scientists who must hold at least a Master's Degree in Science. These 650 scientists are pledged *against* evolution

and *for* the Biblical account of Creation, Adam and Eve, and Noah's Deluge.

- *The Bible-Science Newsletter* of the U.S.A. is producing a monthly publication of scientific facts against evolution.
- *The Institute For Creation Research* is producing literature and is providing highly qualified scientists as debaters, taking the truth to university campuses and public meetings by open debates against evolutionist professors.
- In Australia, *The Creation Science Foundation* has taken over the work of The Evolution Protest Movement, importing books, publishing literature and providing qualified speakers for schools, meetings and seminars. They have many qualified scientists, some of whom have resigned from good teaching positions in order to devote their full time to the crusade against evolution.

The ranks of anti-evolution scientists are growing; but the mass media *ignores them*, or else discredits them by disparagement. We can also quote some giants of science who have rejected evolution outright:

- *Sir Ernst Chain,* F.R.S., Nobel Prize winner for penicillin.
- *Louis Vialleton,* who was Professor of Zoology, Anatomy and Comparative Physiology at Montpelier University, France.
- *Professor Louis Bounoure,* former President of the Biological Society of Strasbourg and Director of the Zoological Museum; became Director of Research at the National Center of Scientific Research in France. Bounoure wrote: "Evolution is a fairytale for grown-ups. This theory has helped nothing in the progress of science. It is useless."
- *Dr. Paul Lemoine,* Past President of the Geological Society of France, and Director of the Museum d'Histoire. An editor of the French Encyclopaedia.
- *Professor W. R. Thompson, F.R.S.* For 30 years Director of the (worldwide) Commonwealth Institute of Biologi-

cal Control, Ottawa, Canada; a biologist of such eminence that he was invited to write a preface to the centenary edition of Darwin's *Origin of Species*. His preface demolished Darwinism gently but completely; but, such was his international status that the preface was published with the centenary edition. A devout Catholic, Thompson wrote devastatingly against evolution until his death in 1972.

* *Sir Ambrose Fleming, M.A., D.Sc., F.R.S.* (Physicist). Was President of the Victoria Institute and Philosophical Society of Great Britain. Inventor of the thermionic valve, which made high-quality radio broadcasting possible. He founded the Evolution Protest Movement.

* *Professor Albert Fleishman*, Zoology and Comparative Anatomy, Earlangen University, Germany. He stated: "The Darwinian theory of descent has not a single fact to confirm it in the realm of nature. It is not the result of scientific research but purely the product of imagination."

* *Professor H. Nilsson,* Genetics, Lund University, a Swedish scientist of world standing.

A recent remarkable development is that quite a few leading evolutionists are publicly acknowledging serious flaws in Darwinism, and in the propositions on which evolution theory has hitherto been based. Yet, they are still holding to belief in evolution of some sort.

An excellent book, *The Neck of the Giraffe* (1981), by Francis Hitching, member of the Royal Archaeological Institute, is an example.

Another splendid book, *Darwin's Enigma* (1984), by Luther Sunderland, is based on the author's interviews with officials of five leading natural history museums: Dr. Colin Patterson (London), Dr. Niles Eldredge (New York City), Dr. David Raup (Chicago), Dr. David Pilbeam (Boston) and Dr. Donald Fisher (New York State). The book ranges far and wide, but it illustrates that the shortcomings of evolution are widely recognized at the top.

All this raises the question: Why do many leading minds still hold to evolution? I think that many brilliant minds have been so molded in established evolution science that there is a blockage which excludes the idea of supernatural Creation. And there is no third alternative.

In the counter-attack against evolutionism, the major work is being done by dedicated non-Catholic Christian scientists, and this small book draws upon their scientific findings.

It seems that the revolution against evolution is mounted mainly by non-Catholics, while Catholics, by and large, have dropped their defenses. Among Catholics, evolutionism is gaining ground because they are not informed about the Church's pronouncements and certainly are ignorant of the recent findings of science. As a result there is spreading among Catholics an evolution-based Modernism, and that specially dangerous brand of Modernism, namely, the evolution theory of the late Father Teilhard de Chardin.

Evolution Infects Christianity

Evolution speculation was an intellectual diversion for centuries. In the early 19th-century it was increasingly active in some circles, but it was not popular until Charles Darwin proposed Natural Selection as a key ingredient. Within 10 years, evolutionism was sweeping through England and the Western world. The theory was formulated and propagated by people who disbelieved, and who even opposed, Biblical Christianity. Christians were caught off balance, ill-prepared to counter the evolution gospel on scientific grounds. In any case, so powerful was the propaganda and so anti-religious was the intellectual attitude, that logical argument would be swept aside by the tide.

The triumph of Darwinism was complete. In time, the minority of scientists who disagreed chose to remain silent rather than arouse futile argument. The propaganda machine steadily persuaded Christians that evolution is unchallengeable. Christians began adapting doctrines and re-interpreting Biblical Creation to fit the ostensible "science" which taught that beasts changed into men over millions of years. The guideline became: Religion must yield to science. Thus, as evolution belief dominated, so did Christian beliefs weaken.

Within the churches there emerged "Christians" who felt that Christianity must be updated to satisfy the new enlightenment—who had lost their faith, but who would not quit the fold. This movement was Liberal Protestantism. It influenced many Catholics, and, around the turn of the century, it gave rise to Modernism in the Catholic Church—Modernism which Pope St. Pius X called "the synthesis of all heresies." (*omnium haeresum conlectum*).

Modernism re-interprets Catholic dogmas and re-casts the whole Catholic system to conform to popular science and the modern outlook. In the words of a Protestant authority, K. Holl (*Der Modernismus*):

> The struggle no longer revolves on an isolated dogma . . . but on the totality of the Christian faith as the Catholic Church has understood and proclaimed it. A group . . . has tried to make, between Catholic faith and modern thought, a reconciliation which would end in reality in the complete overthrow of the whole theological and hierarchical system of Catholicism. (Quoted by Father John McKee in *The Enemy Within the Gate*.)

This could not have happened without the General Theory of Evolution, which is essentially anti-God. Through the greatest propaganda operation of all time, evolutionism is so ingrained in modern thinking that its anti-religious essence is lost sight of. Christians are so misinformed that they are embracing evolutionism with fervor. We are now seeing Christians allied with anti-Christians to promote the ungodly gospel of evolution.

Pius X effectively combatted Modernism early this century. However, in recent years the mass media and the educational system are forcing evolutionism and naturalism into the minds of a whole generation, not just a clerical clique. This has contributed to the resurgence of Modernism on an unprecedented scale. With it has come a withering of the sense of the supernatural; a de-mythologizing of the Bible; disbelief in miracles; confusion of dogmas and doctrines. Jacques Maritian, in *The Peasant of the Garonne*, describes it: "The Modernism of Pius X's time was only a modest hayfever" compared with that of today.

Do I exaggerate?

Firstly, note that the General Theory of Evolution is accepted more or less in many Church schools, and in many seminaries and convents and by many modern theologians. And what is the theory teaching? It is teaching that hydrogen gas evolved into man by purely natural processes.

Secondly, identify precisely the forces of anti-God today. Foremost are Marxism and Secular Humanism. Marxists are openly anti-God. Secular Humanists are more devious; they call themselves "non-theists" to disarm their intended victims. Nevertheless, both have the same unswerving purpose, namely to dethrone God and eradicate Christianity; and their prime target is the Catholic Church. In this purpose, their principal tool is the General Theory of Evolution, which replaces Special Creation and eliminates the personal Creator God.

Thirdly, evolution is the basal doctrine of Marxism (and its creature, Communism) and of Secular Humanism. Their credibility is based on Evolution. They are not viable without Evolution. Discredit Evolution and you topple Marxism, Humanism, and their apostate "Christian" ally, Modernism.

Fourthly, many dedicated scientists have provided us with the scientific case against evolution. Christian churches and church schools have now available the scientific weapon for destroying evolutionism, and thereby paralyzing the enemy.

To use this weapon is a duty for Catholics since Pope Pius XII, in *Humani Generis* (See Appendix A), stipulated that the facts *against* evolution must be given due weight and consideration. If Catholic educators are not fulfilling this duty, then the onus falls on parents.

Revolution by Stealth

Infiltration is the new strategy. If any organized body hinders the march of anti-God, you can bet it will be infiltrated. In a notable book, *Athanasius and the Church of Our Times*, by Dr. Rudolph Graber, Catholic Bishop of Regensburg (Germany), there is a grim passage which I cannot forget: "The point to be noted here above all is the change of strategy which can be dated to about the year 1908: 'The goal is no longer the destruction of the Church but rather to make use of it by infiltrating it.' " Bishop Graber was referring to J. M. Jourdan's *Ecumenism As Seen By A Traditional Freemason*, 1965.

We find another disturbing passage in what Monsignor John McCarthy, President of the Roman Theological Forum, wrote in 1972:

> From outside the Church a plan of subversion has been in effect for over 50 years, implemented by the most powerful subversive organization in the history of mankind . . . Within the Church the number of openly declared Marxist revolutionaries is growing. They are a minority, but a revolutionary element of this kind always is.

He adds that the Marxists are aided by the vaguer leftists in the Church, who have no conscious intent to transform the Church into Socialism. These Liberal Utopian dreamers are dangerous because "they are promoting a revolution whose aims they know not."

From New Zealand comes another warning. Father G. H. Duggan, writing in *The Tablet* (May 6, 1981), discussed the new "liberation theology" which is getting the Church's mission tangled with social revolution; with Marxist priests trying to reconcile their religion with a Marxist dogma that knows no God; and with theologians, Protestant and Catholic, trying to justify the thesis that violence is necessary in the cause of social progress and reform. Father Duggan stresses that religion and Marxism do not reconcile: ". . . the Christian elements eventually evaporate, leaving a residue of pure Marxism . . . and the 'Christian Marxist' becomes indistinguishable from the atheistic variety."

These authorities are sounding warnings about Marxist infiltration. However, if Marxists are adept at penetrating the Christian citadel, humanists and liberal Modernists are in there, too. Homer Duncan, in *Secular Humanism* (1979), informs us how they operate:

> The false teachers in the Christian Churches do not generally call themselves humanists, but are more commonly known as modernists or liberals. Unlike the humanists, most liberals do believe in God: not the God of the Bible, but a god of their own invention and imagination. They deny the supernatural fundamentals of Christianity.

If they are in, how did they do it? How could these antitheses of Christianity enter into and thrive within the Christian stronghold? It seems Christians made it easy for them. Marxism, Humanism and Modernism, all three are fruit of the same root—the General Theory of Evolution. As Christians "progressed" into evolutionism, the fruit seemed less repugnant; and, for some, tempting. Infiltration was not noticed. The ominous three slid into positions of influence.

There remains one burning question: How did the Christian mind embrace the essentially infidel theory of evolution? The answer to that is found in the compulsory education system. Again, Homer Duncan is enlightening:

> This battle is being fought in our public schools and, unknown to most Americans, the humanists have been winning the battle so far in the Twentieth Century. The false evolutionary hypothesis, which has widely been accepted as scientific fact, has all but destroyed the basis for education as it existed at the beginning of the century . . . Now, both Christians and humanists recognize the great impact that evolutionary humanism has made on traditional theism through the public education system.

Actually, there are many comfortable Christians who do not yet recognize the impact of evolutionary humanism; but the humanists certainly do. Homer Duncan tells us how humanist Paul Blanshard recorded his satisfaction with progress in 1978:

> I think the most important factor moving us towards a secular society has been the educational factor . . . The average American child now acquires a high school education, and this militates against Adam and Eve and all other myths of alleged history.

Duncan names the agencies which he contends are promoting humanism in America. There are no surprises in the first few: Atheist Association, Humanist Association, etc.; but some people will be surprised to find that among the prime purveyors of humanism the following are named:

• The United States Government; most powerful and effective, mainly through its control of education.

- The United Nations itself.
- Colleges and universities all over America (both state and denominational schools).

It would be comforting for Europeans and Australians to imagine that the same forces are not reaching beyond the United States. It would also be dangerous naivete. I wonder, is there any non-trendy Christian parent who is not worried by the new complexion of education. We recall the enthusiasm among Australian educators for MACOS (*Man— A Course of Study*) and the fairly lonely voice that blasted it. Dr. Rupert Goodman, Reader in Education at the University of Queensland, warned that MACOS was materialist and humanist in orientation. In featured articles in *The Courier Mail*, Brisbane (Nov. 8 and 9, 1977), he pointed out where MACOS was leading children:

> Children are led to believe that man not only evolved from the lower animals, but the explanation for his social behavior is to be found mainly in his cultural environment . . . MACOS appeals strongly to the evolutionaries and to the secular humanists, but these are not the values which should underlie our school system.

Nevertheless, the generality of educators set great store on MACOS, and they would mostly be sincere and well-meaning teachers. They would also be products of the system, evolutionary-humanist public education.

There are splendid teachers who do not like the trends, but their voices become more lonely—and even silent. Meanwhile, the education machinery is tooled toward producing what Homer Duncan described, and those it produces are grasping more of the control levers. The products of the system in good conscience propagate the system. So will the system empty God the Creator out of the student mind, leaving it vulnerable to humanism and Modernism, and even to Marxism—because a godless mind is an open field to Marxism.

Darwin's Evolution gave to Lucifer the perfect weapon with which to shake the foundations of Christianity. Man was given an alternative. He could choose between Creation and evolution.

We see the result in today's *secular man*. Heedless of any Creator God, he acknowledges no Commandments from a Creator. Thus is removed the source of *authority*, and lost is the sense of *moral absolutes*. Gone is the concept of rendering a *final account* to an almighty God. It is small wonder that *all* authority is breaking down.

Modern man views the awesome universe, not as a hymn to the Creator, but as patterns of matter blindly shaped by chance. Man is taught that mankind is but part of the vast evolutionary process, but is the summit of the process. So, Man is his own "god." There is nothing above Man which can decree "Thou shalt . . ." or "Thou shalt not . . ." You may love your neighbor or you may mug your neighbor. What's the difference? Both of you are merely assemblages of atoms—and atoms have no conscience or rights.

So much for the world outside. What of the Church?

Lucifer is clever. He knows that by dislodging one stone (Original Sin) he can collapse the Christian structure. But, to dislodge Original Sin he must get rid of Adam. Adam must go; and the ape-men take his place. The great channels of information tirelessly proclaim that everything evolved, and that apelike animals turned into men, but *not* into *one* splendid first man and *one* superb first woman. They tirelessly proclaim the reverse, namely, that evolution would have produced *many* first humans, *groups* of them, *populations* of brutish first humans who were little better than their animal parents.

The message is being drummed into young and old: Adam

is a myth. Adam was a tribe. Adam is a symbol for a population of first humans.

That is polygenism—many Adams, many first men—and its results are devastating. It is at the base of the errors which afflict Christians today.

Pope Pius XII ruled against polygenism in *Humani Generis* in 1950; but many modern theologians are performing prodigies of polemics to admit polygenism and evade the Papal ruling.

Pope Pius XII also stipulated that the facts against (as well as those for) evolution must be properly weighed and adjudged. Yet, Church schools are producing a generation of evolutionists who have never heard a single fact against evolution.

We see Lucifer's consummate strategy. Evolution has deadened man's thoughts of the Creator and his sense of the supernatural, as well as his trust in the Bible. Erosion of the Bible began with Adam and has spread even to the New Testament.

The trump card was polygenism. This plays havoc with the central dogma of Original Sin. When Original Sin is discredited, all the dogmas start to fall like dominoes. Without Original Sin:

- *Baptism* loses its traditional meaning;
- *Redemption* (from the effects of Adam's sin) is confused;
- *The Immaculate Conception* becomes meaningless;
- *Papal Infallibility* is open to challenge, because a Pope infallibly defined that Mary was conceived free from Original Sin;
- *Personal sin* loses credence, and is now widely disregarded.

One by one the dogmas have been emptied of meaning, and now, under the naturalism of Evolution:

- *The Virgin Birth of Christ* is questioned;

- *Matter and Spirit* are regarded as the same;
- *Miracles* are denied, even the *special creation of the soul.*

The tragic waves have spread, overturning all Catholic doctrines; disclaiming the spirit world; renouncing Heaven, Hell, grace, the Cross, angels, devils. Even God somehow is made a part of the grand sweep of evolution. Lucifer knew that, as evolutionism advanced, Christianity would recede.

Teilhard de Chardin and The New Religion

The general theory of Evolution is diametrically opposed to Christian revelation and creed. It opened a chasm between modern thinking and traditional Christianity. Ostensibly to bridge this chasm, and professedly to clothe Christianity in a garb acceptable to science, there came a Jesuit priest, Father Teilhard de Chardin. Whatever his personal motives may have been, his ideas have done more damage to orthodox Catholicism than those of probably any other person in history. His "evolution-theology" has raised a new religion beside the old one. There are now two religions called "Catholic," with a lot of confused Catholics in between.

Teilhard gained a reputation in scientific circles for his part in the setting up of the Piltdown Man (now discredited) and Peking Man, the real story of which is tainted with equally discreditable procedures. These activities are discussed later in this book.

Teilhard's mind was firmly locked into evolutionism on a grand scale. He proclaimed: "Evolution is not just a hypothesis or theory . . . It is a general condition to which all theories, all hypotheses, all systems must bow and which they must satisfy if they are thinkable and true." To Teilhard, evolution and polygenism were the essential realities which Christianity must perforce satisfy.

In 1922, he wrote an essay which treated Original Sin in

a way contrary to Church teaching. By mistake it went to
the Vatican, and Teilhard was nearly excommunicated. He
was forbidden to teach or preach; but he wrote secretly, and
his pamphlets were passed from hand to hand. He wrote
several books formulating a Christianity which bowed to
total evolutionism. His books were refused a Church *Imprimatur* and remained unpublished.

Bridges: (a) His followers claim that Teilhard built a
bridge between religion and science. As regards the religious end of the bridge, a respected theologian, Cardinal
Journet, described Teilhard's work as "Disastrous! . . . It
contradicts Christianity." Even more importantly, the official Catholic Church has warned against Teilhard's evolution theology in several pronouncements and actions. (See
Appendix B.)

As regards the scientific end, it is hard to imagine any
scientist using Teilhard's bridge to approach religion. England's famous man of medicine, Nobel Prize winner, Sir
Peter Medawar, stated that Teilhard's works lack scientific
structure and that his competence in the field of science is
modest. In *The Art of the Soluble* (1967), Sir Peter dismissed
Teilhard's works as a bag of tricks for gullible people—for
people whose education has outstripped their capacity for
analysis.

(b) Teilhard's work is also claimed to be a bridge between
Christians and Marxists. Dietrich von Hildebrand (in *Tro-
jan Horse in the City of God*) quotes Teilhard's own words:
"As I love to say, the synthesis of the Christian God (of the
above) and the Marxist God (of the forward)—behold! that
is the only God whom henceforth we can adore in spirit
and in truth." Commenting on this, von Hildebrand says:
"In this sentence the abyss separating Teilhard from Christianity is manifest in every word."

The non-Catholic biologist, Bolton Davidheiser, Ph.D. (in
Evolution and Christian Faith) tells us:

The delegates of the Twentieth Annual convention of the American Scientific Affiliation were told that "in Europe, both Christians and Marxists find his thought the most helpful bridge this century offers between what once seemed their irreducibly opposing views."

Further to these references to Marxism, it is noteworthy that Pope Pius XII in *Humani Generis*, without mentioning Teilhard, drew attention to extreme evolutionists whose monistic or pantheistic speculations are eagerly welcomed by the Communists as being powerful weapons for popularizing dialectical materialism.

Pantheism (?): In a letter, January 26th, 1936, Teilhard wrote:

What increasingly dominates my interest . . . is the effort to establish within myself, and to diffuse around me, a new religion (let's call it an improved Christianity if you like) whose *personal God* is no longer the great neolithic landowner of times gone by, but *the soul of the world* . . . [emphasis added].

Matter and Spirit: Essential to Teilhard's whole system is the assertion that matter and spirit are one. He uses the Spinozan idea that matter has a "within" and a "without." From the outside it is matter; but, looked at from within, this matter has consciousness and thought. Also, the "within" and the "without" are developing in complexity.

Teilhard taught that primitive particles of matter assembled into more complex arrangements until some most complex arrangements burst into life. Lifeless matter had become alive, and it continued to complexify until it reached a "boiling point," whereupon the living matter became conscious.

The animal stage had been reached. The complexifying continued. The brains of some higher animals attained such complexity that, in one type of animal, thought was generated and the animal became man. Matter, in the shape of man, had begun to think.

From that point, Teilhard proposes that evolution is sweeping man's thinking-consciousness upwards toward the climax when all humanity will merge into a "super-consciousness" with common thought and common will. He calls this the Omega Point where, he says, all creation will be united with Christ (the Cosmic Christ, evolutor of the world) and absorbed in God.

To claim that matter and spirit are the same leads to denial of the spirit world followed by rejection of the supernatural character of Christianity. I detect an element of cheating in the proposition that the material and the spiritual are one. It is as if Teilhard saw that he faced a problem in getting mind to evolve from matter, and he got over the problem by pronouncing in advance that mind and matter are the same substance. His disciples gravely nod in agreement, not because Teilhard produces evidence, or even a good argument, but simply because Teilhard says so.

The "Cosmic Christ": "Christ saves. But must we not hasten to add that Christ, too, is saved by evolution?" That is another gem by Teilhard.

Jacques Maritain's reaction was that Teilhard is most anxious to preserve Christ; but "What a Christ!" This is no longer Jesus, the God-Man, the Redeemer; this is the initiator of a purely natural evolutionary process, and also its end—the Christ-Omega. Any unprejudiced mind must ask: Why should this cosmic force be called Christ? Teilhard has dreamed up an alleged cosmogenic force and has then tied onto it the label "Christ." Maritain warns that we must not be fooled by this subterfuge of wrapping pantheism in traditional Catholic terms. He explains why: Teilhard, the obsessed evolutionist, has a basic conception of the world which cannot admit traditional Original Sin. Consequently his world has no place for the Jesus Christ of the Gospels, because, without Original Sin, the redemption of man through Christ loses its inner meaning. (*The Peasant of the Garonne* by Jacques Maritain).

Teilhardism Invades: Teilhard de Chardin died in 1955. Thereupon, a group of people who were extreme evolutionists, and some of whom were atheists, had his works published without the authority of his Jesuit superiors. From that moment, Teilhardism invaded the Catholic Church on a large scale. Teilhard's ideas entered modern catechetics. Children whose parents were unaware of Teilhard de Chardin were indirectly subjected to his ideas.

It has been said that the real danger to the Church is Modernism and that evolutionism is only a minor academic exercise. Such a view misses the point that Modernism and Teilhardism have their source and lifeblood in the General Theory of Evolution. Logic, theology and sweet reason usually will bounce off the Modernist. However, if you discredit evolution, you collapse the foundation of it all and the Modernist is left without support. While this might not cause a change of heart in a dedicated Modernist, it should fortify the ordinary person against the intellectual seduction of Modernism. Above all, if we can get through to our young people that evolution is unscientific nonsense, they will be spared the religious doubts and compromises which propel them into the pseudo-sanctuary of Modernism and Teilhardism.

Chapter 2

THE DEMISE OF DARWINISM—
OCTOBER, 1980

Darwin's *Origin of Species* was published late in 1859. Twenty-two years later, its triumph was so complete that G. J. Romanes wrote in *Nature*: "And, never in the history of thought, has a change been effected of a comparable order of magnitude."

Darwin proposed that complex plants, animals and man evolved from primitive forms of life by a process of gradual change and natural selection. This idea, aided by rich and powerful forces, bulldozed through all opposition, brooked no argument, and has become the world view. Textbooks proclaim it; generations of students have been taught it.

Then, remarkably, in October 1980, the bulldozer made strange noises. It actually rolled backwards. At an historic conference in Chicago, 160 of the world's leading evolution experts faced the facts of the fossil record and virtually pronounced the death of Darwinism. The experts admitted that 120 years of digging up the fossil record has shown that there are no fossil links between one species and another.

In a report of the conference, *Newsweek* (3/11/80) stated:

> The missing link between man and apes . . . is merely the most glamorous of a whole hierarchy of phantom creatures. In the fossil record missing links are the rule . . . The more scientists have searched

19

for the transitional forms that lie between species, the more they have been frustrated.

Newsweek further stated:

> Evidence from the fossil record now points over-whelmingly away from the classical Darwinism which most Americans learned in high school; that new species evolve out of existing ones by the gradual accumulation of small changes, each of which helps the organism survive and compete in the environment.

Unfortunately, although the experts have confessed that Darwinism is wrong, they are not advising us to go back to the beliefs that Darwinism un-taught. In the media reports of the Chicago conference, we find no contrition for so long misleading the world. Instead, evolution will be salvaged at any cost. Having admitted that the evidence for evolution is missing, these experts are now working on a theory of evolution without evidence. They are suggesting evolution by huge jumps which would leave no fossil evidence.

Fossil evidence shows that species are stable. To get around the evidence, it is now proposed that individual species remain unchanged for long periods, but, occasion-ally, a hypothetical mutation produces a monster, something different from the parent stock, which (contrary to all genetic experience) happens to survive and produce offspring which are a new species. It means long periods of no change (equilibrium) punctuated by sudden large genetic change. So, the new theory is called evolution by "punctuated equi-libria."

Actually the Chicago conference performed a somersault. The experts are now embracing Professor Goldschmidt's the-ory of the "hopeful monster" which, for 40 years, they derided as unscientific. In their hearts these scientists must know that the "hopeful monster" or "punctuated equilibria" or "quantum leaps"—call it what you like—violates all relevant laws of science. By adopting it to prop up evolution, they admit that evolution is scientifically bankrupt.

The *Newsweek* report says:

> The new theory also raises the troubling question of whether man himself is less a product of 3 billion years of competition than a quantum leap into the dark—just another hopeful monster whose star was more benevolent than most.

Darwin contributed the idea of *natural selection*. It was a winner that brought a flood of converts to evolutionism. Since the Chicago conference, natural selection becomes relegated to a role of little importance.

Ever since Darwin, the expectation of finding transitional fossils has been the essence of the theory of evolution. When evolution's own leaders admit: NO transitional fossils! they themselves have forfeited the game. However, we shall still present the case against evolution, even if only to counteract teachers who never heard about the Chicago Convention of 1980.

The Second Law of Thermodynamics

"If your theory is found to be against the Second Law of Thermodynamics I can give you no hope." (Sir Arthur Eddington, astronomer).

The most fatal objection to the theory of evolution is that it goes against the Second Law of Thermodynamics.

This Law can be stated in various ways. For our purpose it means:

(a) Natural processes always tend toward disorder; they move from orderliness to disorderliness.

(b) The simple will never produce the more complex.

It means that the universe is running down; that all natural systems are degenerating from order to disorder. Natural systems tend toward increasing *entropy* (which means increasing disorder).

Evolution requires the opposite. It requires the universe to run uphill. It requires random molecules to assemble themselves into organized and increasingly complex systems. Evolution requires the simple to bring forth the complex. It requires jellyfish to transform into humans. Even more than this, it requires hydrogen gas to evolve into thinking man by purely natural processes.

Living Systems (plants and animals) are extremely complex and highly organized arrangements of matter. How could they arise from disordered molecules of matter? The Second Law rules that order cannot spontaneously arise from disorder, nor can the complex arise from the simple.

Evolutionists who will not admit a deity must offer a natural explanation. They must explain how living systems do not break a Law that is admitted to be unbreakable. They propose that energy from the sun performs the marvel. They say that the sun's energy "pays the entropy debt" required for mustering molecules into living systems as well as engineering their evolution ever upwards.

Something more than raw energy is required, namely energy guided by intelligence, be it the intelligence of the engineer, the craftsman or the deity. Implanted in the seed and in the egg are special mechanisms such as photosynthesis and DNA blueprint. These are of an intricacy beyond human expertise, an intricacy that could have been fashioned only by a greater than human intelligence. They provide the intelligence which guides and transforms the sun's raw energy into building life systems. The creationist sees this intelligence as that which wrote the Second Law.

The General Theory of Evolution is teaching that primitive gas evolved into humans by natural processes. It postulates innumerable intermediate stages and enormous time spans; but these are merely distractions from the essential question of thermodynamics: could it happen? From gas to man would involve such staggering decrease in entropy (disorder) as to mock the Second Law of Thermodynamics. Therefore the scientific answer to the question can only be: it could *not* happen.

Darwin's hypothesis broke the Second Law of Thermodynamics, because Darwin was saying that the simple could produce the complex.

Darwin had been taught at Cambridge that species never vary. He visited the Galapagos Islands and there he saw variation within certain species. He overreacted. He extrapolated observed variation into the unobserved, into imaginary variation across barriers of species, into a hypothetical world of evolution of lower kinds into higher kinds.

By dropping the barriers between species (or kinds) Darwin opened a dream world in which jellyfish can become elephants, and a sea squirt can end up as a Shakespeare. The world welcomed the dream. Evolution is the only alternative to Creation. Evolution suited the mood of the nineteenth century and of the twentieth century, seeking an alternative to a Creator God.

There is no argument about variation. The argument is about the *limits* of variation. Variation within limits has produced races of men, breeds of rabbits, varieties of pigeons. Darwin supposed variations were by chance. Since his day, the science of genetics has revealed that microscopic genes in the living cell control the variations of individuals within the species or the kind. The genes of a butterfly or an ape or a fowl carry the blueprint which determines the appearance and the character of the butterfly or the ape or the fowl. The genetic blueprint allows great variety within a species or kind, but no further. There is a genetic barrier which prevents change beyond that. This barrier confronts every animal and plant breeder. Always a limit is reached beyond which no further improvement of the line can be obtained.

Is there, then, any genetic process that can break the barrier? All genetic research says: NO! However, the neo-Darwinists will answer: YES! Mutations of genes! Mutations and Natural Selection! Therein lies the heart of the case of the neo-Darwinists: (1) Mutations and (2) Natural Selection. Let us look at both.

Mutations. The genes are very precisely arranged in the DNA in the living cell. An organism reproduces itself by a complex process which is so exquisitely engineered it suggests the miraculous. The finite facts are basically that, when sperm impregnates ovum, the DNA produces a replica of itself which is passed onto the new cell. The process is almost mistake-proof and the DNA is faithfully replicated and handed on generation after generation.

But occasionally, rarely, a mistake occurs which is called a mutation (or change). The mistake alters the correct DNA blueprint. The mistake becomes part of the genetic pattern of the recipient and becomes inheritable by future descendants. Being a mistake, it would be unlikely to upgrade those affected. In fact, for practical purposes, mutations are always harmful or useless, or even fatal. Fortunately, mutations are rare. They occur in approximately one cell out of 100-million, though estimates vary.

Modern Darwinists have seized on mutations to provide the mechanism for upward evolution. They go further and, with some sort of logic, they claim that the effect of mutations would appear in groups or populations. Applying this to man, they say that the gene pool of whole groups of "hominids" (man-like apes) would have been affected by mutations. This would have produced groups of first humans. Instead of one first human couple (Adam and Eve), there would have been groups of low-type first men and first women, born of groups of non-human parents. This is Polygenism (many Adams), a theory which has had a disastrous effect on people's faith in the Christian religion because it appears to undermine the first books of the Bible and to disturb the central Christian doctrine of Original Sin.

Mutations must be recognized for what they are, namely mistakes that are damaging, not constructive. In fact, all forms of life have wonderful *repair mechanisms* to guard the DNA against the ravages of mutations—repair mechanisms that are fundamental to the survival of living organisms. However, some mutations slip through these defense mechanisms with usually undesirable results. The results of mutation are diabetes, club feet, hemophilia, mongolism, color blindness, sickle-cell anemia, creeper chickens, calves with deformed jaws, fruit flies with crumpled wings or no wings, seedless oranges (not viable in the wild), and so on and so on.

The hopelessness of mutations to produce evolution was confirmed by classic experiments on the fruit fly. Fruit flies breed rapidly. Furthermore, they were given doses of radiation which speeded up their mutation rate by 15,000 percent. After a long experiment involving 25-million fruit flies, they refused to turn into anything else. Certainly there was plenty of variation: stunted wings, lack of wings, yellow eyes, useless eyes, abnormal feet and bodies. There was grotesqueness; there were freaks. Perhaps the strangest was an apparent foot instead of a proboscus. But it was a fruit fly foot, not a bull's foot. And they were fruit fly wings and bodies, though deformed. Never was there the start of a new organ of a different species. And, most important: No matter how monstrous the offspring, it was able to breed with the parent stock, if it was capable of breeding at all. This meant that it remained the same fruit fly species. Mutations can do all sorts of things to the organs of a species, but never produce a new organ. For example, a mutation can produce a baby with deformed arms, but that is a different thing from producing a baby with wings, or a baby with wheels.

Evolution's case has been based on a claim that there

occasionally happens a "good" mutation, preserved by natural selection; and that an accumulation of these provides "new genetic information" for upward evolution. Even if one in 1,000 or even one in 100 is a "good" mutation, when we consider the rarity of mutations and the damage of the bad mutations, it is a hopeless base on which to hypothesize upward evolution. What is more, there is no documented example of a good mutation. "Good" mutations are as hypothetical as the whole evolutionary idea.

Mutations are misfortunes. They could not generate evolution. Many good scientists admit this. I cannot go much higher than Sir Peter Medawar, F.R.S., and Nobel Prize winner. In his book, *The Art of the Soluble* (1967), he frankly admits that, at present, science knows of no genetic process that could produce variations required for evolution. And he says that "It is not enough to say that 'mutation' is ultimately the source of all genetic diversity, for that is merely to give the phenomenon a name . . ." Sir Peter indicated that "What we want, and are very slowly beginning to get, is a comprehensive theory of the forms in which new genetical information comes into being."

In the years since Sir Peter wrote with hope, have they found that comprehensive theory? The answer is, not only have they not found it, they are so far from finding it that the hope is now dead. This is evidenced by the admissions made by so many leading evolutionists at the historic Chicago Convention in October, 1980, already referred to. Reports of the convention can be found in *Science* (Nov. 21, 1980: "Evolutionary Theory Under Fire") and *Newsweek* (Nov. 3, 1980: "Is Man a Subtle Accident?")

The speeches contain much speculative rhetoric; and the only substance seems to be a general admission that the fossil record has failed to produce transitional forms, and therefore it is time to abandon the neo-Darwinist idea of evolution by small changes through small mutations preserved by natural selection. Instead, a new theory gained support, namely, that species evolved by sudden leaps, which would leave no fossil evidence and which need not confer a survival advantage. This is the reverse of Darwinism. Though called "punctuated equilibria," it is really a desperate acceptance of the long-derided "hopeful monster" theory.

What is the Hopeful Monster Theory? Dr. Richard B. Goldschmidt, a geneticist of world rank and a dedicated evolutionist, conducted experiments for 25 years with gypsy moths. He found that they would not transform into anything else, so he decided that there was something wrong with Darwinism. He also noted that, after a century of laying bare the fossil record, there had been found not one transitional fossil to support Darwinism.

To his credit, Dr. Goldschmidt challenged his evolutionist colleagues to discard Darwin's evolution by small changes. This meant that Goldschmidt had to offer an acceptable alternative. To his discredit, he disdained the Creation alternative. In 1940, he proposed his "hopeful monster" alternative. He pointed out that monsters are sometimes born (a sheep with only two legs; a calf with two heads) and, being hopeless monsters, they die. But, suppose that occasionally one lived, and that this hopeful monster bred and transmitted its peculiar genes to descendants; it might be the bridge between species "A" and species "B". Then a different monster might bridge species "C" and "D"; and other monsters would bridge all the gaps between all the major kinds. Of course, none of them would leave any fossil clues of the bridgings.

In this radical theory he concurred with the suggestion that this could mean that a dinosaur laid an egg and a bird hatched out of the egg.

While Goldschmidt challenged his colleagues to produce evidence for their neo-Darwinism, they were content to deride his "hopeful monster" theory as a scientific joke. Finally the facts had to be faced. A noted evolutionist, Dr.

Stephen Jay Gould of Harvard, in 1977 wrote in the American Museum's monthly magazine, *Natural History*, a column titled, "The Return of the Hopeful Monster." It was shattering. It predicted that Goldschmidt's no-evidence theory would have to be embraced to some extent in order to meet the absence of evidence for evolution.

Three years later, at the convention in Chicago of 160 leading evolutionists, Gould and Niles Eldredge (of the American Museum of Natural History) were the leading proponents of the theory of "punctuated equilibria," or evolution by occasional huge jumps, which is essentially the "hopeful monster" theory. Although there was much verbal jostling, and although some had reservations, it seems that "the majority of the 160 of the world's top paleontologists, anatomists, evolutionary geneticists and developmental biologists supported some form of this new theory of 'punctuated equilibria.' " (*Newsweek*).

One can imagine Goldschmidt's ghost of a grin.

Hopeful Monster?

Natural Selection. Darwinism regarded Natural Selection as the key to evolution. At that Chicago convention in 1980, the world's leading evolutionists demoted Natural Selection to a very minor role. However, even when Natural Selection was being hailed as the wonder-worker of evolution, creationist scientists were pointing out that the true task of Natural Selection is to conserve the quality of a species, not to transform it into a different species.

Darwin based his theory on his famous (and false) analogy which said: If man, using artificial selection, can breed better animals, surely nature, using natural selection, can do likewise and on a bigger and better scale without limit.

Analogies are risky. Darwin's analogy ignored the stringent rules of artificial selection:

(a) Mating must be restricted to selected individuals;

(b) The selected stock must be isolated and protected.

Natural selection cannot apply these rules in the wild.

Artificial selection has shown the limits of variability. Breeding experts have obtained rapid improvement in selected stock for a few generations. Then a point is reached where no further improvement can be obtained along that line. This shows that there are genetic barriers around each kind. Emphatically, it proves that one kind will not breed into a different kind.

Predation: It was also claimed that natural selection worked by predation. Predators should weed out inferior prey. This should upgrade the prey species. It should also upgrade the predator species; only the better predators could catch the upgraded prey. However, the evidence indicates that inferior prey are avoided by predators if possible. Many predators, especially carnivorous mammals, prefer to hunt healthy prey, even when weaker ones are available. Overall, it is mostly chance or luck which determines which animal (or insect or bird or plant) is eaten.

An evolutionary mechanism must add to the gene pool,

and this cannot be done by predation, especially when chance governs the predation.

Natural selection is a real and operating process, but it does not create nor upgrade. It is a conservation process; it conserves the status quo. In particular it counteracts bad mutations and kills off inferior mutants. It also keeps population numbers in check when they exceed a proper balance with the environment.

A theory of evolution based on good mutations and natural selection crumbles when exposed to the light of true science.

(Interesting articles on Natural Selection are in *Creation Research Society Quarterly* magazines of September, 1976, by E. Norbert Smith, Ph.D.; December, 1976, by William J. Tinkle, Ph.D.; and September, 1979, by Randall R. Hedtke.)

Embryology: The story is still told that the human embryo has *gill slits* like a fish and the *tail* of an animal. The story began when Darwin's zealous ally, Ernst Haeckel, propounded his "biogenetic law." This "law" asserted that the embryo re-enacts the stages of its past evolution, or "the developing embryo climbs its ancestral tree." Haeckel faked photographs and falsified drawings in order to provide evidence for his "law." Darwin enthusiastically declared that Haeckel's biogenetic law was "evidence second to none" in favor of evolution.

This "biogenetic law" dominated biology for many years until the new science of genetics revealed its absurdity. It was then discredited. Science threw it out 50 years ago; yet, we still find it being quoted even by some scientists and teachers, and by the media. A typical example of this appeared in a science strip which appeared in an Australian newspaper. It assured the reader that "After a million years, man is still evolving, while still retaining traces of his animal ancestors, including fine hairs on the skin, and, in the embryo stage, a tail and gill slits like a fish."

First, the *gill slits*: The illustration of the human embryo shows them near the head of the embryo. The fact is they are not slits of any sort. They are the visceral arches. The curvature of the neck emphasizes the furrows between the arches. The arches carry blood vessels which carry blood to the head and the back. All vertebrate embryos develop in this functional way up to a certain stage, but the arches and furrows are not slits and have no respiratory function. Future developments for land vertebrates are completely different from those of fish.

Next, the business about a *tail*: To put it as simply as possible, in the embryo the budding legs need plenty of blood. They would not get enough blood if they were at the extreme end of the embryo body. In the illustration you see the leg bud (arrowed). Note that it is at a point a little distant from the end of the notochord which becomes the backbone. Some evolutionists still point to the part of the notochord which extends past the leg buds and unblushingly call it a tail, even though science threw out the idea 50 years ago.

As the embryo grows, its body absorbs the supposed tail except the last four vertebrae. They fuse into the coccyx bone. The evolutionist will now point to the coccyx and call it a vestigial tail, that is, a useless remnant from man's animal past.

The coccyx is certainly not useless. Without a coccyx we could not sit down comfortably. Even more importantly, the coccyx is an anchor post. Anchored to it are ligaments and muscles which control the anus. The coccyx is a vital part of the mechanism of waste elimination.

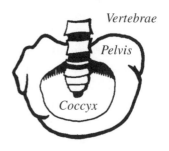

Vestigial Organs. At one time evolutionists had a long list of supposed vestigial organs in man, withered memorials of man's animal past. As medical knowledge increased, it was found that these supposedly useless relics are functional and even vital. A good example is the coccyx which we have just discussed. In the light of modern knowledge, it is hard to understand how a modern science strip could claim that Man retains "traces of his animal ancestors including *fine hairs on the skin*." It is now known that these hairs are active and useful. Each body hair grows in a follicle in the skin. Into the follicle, the duct of a sebaceous gland opens. The glands produce a lubricating fluid which keeps the skin smooth and soft. The flow of the fluid is regulated by the hairs. The hairs have muscles attached to them. When the hair is moved the muscles squeeze out the fluid. The hairs also serve to keep the opening of the ducts free.

The important thyroid gland used to be called vestigial. The appendix and the tonsils were regarded as vestigial; but the modern medical view is that both organs help protect the body against infection, especially when young.

The thymus gland is a classic example. It was considered a purposeless vestige. Recently the thymus has been found to be the master gland which protects the body against infectious diseases. (The fascinating story appeared in the *Reader's Digest*, February, 1967: "The 'Useless' Gland that Guards Our Health.")

Evolutionists used to cite about 180 "vestigial organs" in man and the higher animals. One by one they were struck off the list as their useful functions were discovered. Today there are practically none remaining on the list. Nevertheless, as we have seen above, some diehard evolutionists still cite vestigial organs and mislead the unwary.

Finally, we should note that occasionally a baby is born with a tail-like appendage. Evolutionists tend to claim that this is a throwback to when man's ancestors had tails. The fact is, such a deformity—which is rare—is usually a type of fatty tumor; and it certainly has no relationship to the tail of a monkey. (See *The Creation Explanation* by R. E. Kofahl and K. L. Segraves, 1975).

Chapter 3

THE FOSSIL RECORD

Until the nineteenth century it was generally accepted that the layers of rocks had been laid down by the Deluge. In 1815, a British canal engineer noted that certain rock strata always contained certain fossils. A system began of classifying rocks by their fossils. So far, creationists had no objection; but soon the classification was linked to an evolutionary time scale. The layers of rocks were supposed to represent divisions of time, great ages of time, indexed by their particular fossils. Thus, evolutionists themselves vested their theory of evolution in the fossil evidence. Evolution must stand or fall on the fossil record.

If life began with simple forms and steadily progressed upwards into more and more complex creatures to culminate in man, the fossils should record the progress step by step. If creatures evolved, there would have lived and died and been fossilized innumerable intermediate creatures, steadily bridging the transitions between one kind and the next kind. In fact, there would have been so many intermediates, in successive stages of transition, that we should now have difficulty finding fossils of the perfected kinds amid the overwhelming profusion of transitional fossils. Evolution demands that sort of fossil record; and the Darwinists expected to find it. They have been digging for 120 years. Innumerable fossils have been unearthed, but not one fossil of an intermediate creature. There is not a single transitional fossil to bridge the gaps between perfected kinds.

Then we note another remarkable thing. We find that every

THE GEOLOGIC COLUMN (THROUGH EVOLUTIONIST EYES)			
ERAS	**PERIODS**	**BEGAN** MILLIONS OF YEARS AGO	**EVOLVING LIFE**
CENOZOIC	PLEISTOCENE	1	MAN
	PLIOCENE		HOMINIDS
	MIOCENE		APES
	OLIGOCENE	60	MONKEYS
	EOCENE		
MESOZOIC	CRETACEOUS	130	PRIMATES WHALES MARSUPIALS MAMMALS
	JURASSIC	180	FLOWERING PLANTS BIRDS
	TRIASSIC	220	DINOSAURS CONIFERS
PALAEOZOIC	*CARBONIFEROUS* PERMEAN	280	REPTILES
	PENNSYLVANIAN	310	PRIMITIVE REPTILES SWAMP FORESTS (COAL)
	MISSISSIPIAN	350	AMPHIBIANS
	DEVONIAN	400	SHARKS BONY FISH FERNS
	SILURIAN	450	INSECTS MOSSES
	ORDOVICIAN	500	JAWLESS FISH JELLY FISH
	CAMBRIAN	600	MOLLUSCS SPONGES INVERTEBRATES
PROTEROZOIC **ARCHAEOZOIC**	PRECAMBRIAN	3000	BACTERIA ALGAE

kind alive today, which appears in the fossil record, appears in fossil form similar to its present living form, unchanged, un-evolved.

The fossil record has proved hostile to the very theory it was expected to prove. What the fossil record does show is this: perfected kinds separated by unbridgeable gaps. It is striking evidence for the Genesis revelation that creatures were created according to their kind.

According to the fossils in the rocks, life began suddenly in the Cambrian period, supposedly about 600 million years ago. The rocks older than the Cambrian show practically no signs of life, nothing except a few traces of what may be single cells and algae. There should be an immense fossil record of earlier, evolving life for a thousand million years or more, leading up to the explosion of life in the Cambrian period. Evolution depends on a great pre-Cambrian fossil record and it just cannot be found, except for those few traces of protozoa and algae. Except for these, the pre-Cambrian rocks are barren and lifeless.

But in the Cambrian rocks the fossils appear suddenly. Suddenly there is teeming life. And these are highly specialized creatures, sharply divided into species, genera, families and the rest. The first fossils are quite disastrous for the theory of evolution, and Darwin admitted that this could be an objection to his theory. Indeed, as Douglas Dewar, F.Z.S., exclaimed: "The rocks cry out 'Creation!' "

After the Cambrian period, new kinds appear in the rocks from time to time. But, every time a new kind appears, it appears suddenly, fully perfect, with no link to any previous kind. There are no fossils of transitional creatures.

Dinosaurs are the artists' symbols of evolution. But dinosaurs contradict evolution. The great dinosaurs appeared from nowhere. They flourished for a period. Then they abruptly disappeared, so abruptly that evolutionists are baffled. That is not evolution.

Tyrannosaurus Rex was the king. He was 50 feet long, 20 feet high and had great 6-inch teeth. He was built to giant scale, except for his little arms, so short they would not reach his mouth. Even bigger were the long-necks. *Diplodocus* was 90 feet long. You would need three semi-trailers to spread him out. His feet supported 30 tons. *Brachiosaurus* was tremendous. Even if you climbed a tree 40 feet high, Brachiosaurus could have lifted up his head and looked you in the eye at 40 feet. His feet carried 50 tons. But these colossi were overshadowed when James A. Jensen found remains of a dinosaur, superficially resembling Brachiosaurus, which weighed an estimated 80 tons. To top that one, in 1979, *Science News* reported that Jensen had found an even larger dinosaur. These latest discoveries mean that there walked the earth dinosaurs the size of the blue whale. At the other end of the scale were some midgets, like *Compsognathus*, about the size of a bantam rooster.

Tyrannosaurus Rex

Most dinosaurs were colossal, and they left colossal fossils which could not be missed. But there are no semi-colossal fossils of anything evolving into these giants.

And there is a worse problem: the *Pterosaurs*. These were reptiles; but there is an infinite gulf separating them from any other reptiles. For these were flying reptiles with great, leathery wings. They came in various sizes; the smallest was no bigger than a sparrow. The largest was thought to be *Pteranodon* with a wingspan of 27 feet, until Douglas Lawson, in 1971-75, discovered in Texas, fossils of three pterosaurs, the biggest having an estimated wingspan of 51 feet. This flying leviathan was more than four times the size of our largest living bird, the albatross. It was a flying reptile whose wings exceeded the mere 43-foot wingspan of the F-15A jet-fighter plane.

Pterosaurs

The flying reptiles have no ancestors. The first specimen was a perfect pterosaur. Whence did they come and whither did they go? They are such a problem that one evolution book candidly admits: "All in all, we should have many fewer problems if the pterosaurs had never existed . . ." (*Prehistoric Animals* by Barry Cox, Ph.D., Hamlyn, 1969.)

Among the dead bones of the past, we find no fossil links, no evolution. Then, if we look at the living world of the present, again we find no intermediates between living kinds. Living creatures prove that kinds do not change, no matter how long the time span. For example, using the evolutionists' supposed ages, there is a dragon-fly species still with us after sixty million years. The Australian lung-fish should have evolution in its blood, but it has not changed in 220 million years. Spiders remain unchanged after 300 million years, cockroaches and silverfish unchanged in 350 million years.

Turtles have been turtles for 250 million years of evolutionary time. Turtles have an incredible skeleton. They live inside boxes, and their limb girdles are inside the rib cage. That should mean a lot of evolving. We should find millions of quarter-turtles, then half-turtles, and so on. But we do not. The very first turtles were perfect turtles. There is no fossil of something pre-turtle, nearly turtle.

The hard fact is that every kind of creature living today which appears in the fossil record appears there in form similar to its present form. This was dramatically confirmed by the *Coelacanth* fish. This fish, through one of its relatives, was credited as being the ancestor of amphibians, a vigorous evolutor. It was regarded as extinct for 70 million years. But in 1939, a fisherman hauled up a Coelacanth very much alive. To an evolutionist, this was just as upsetting as if a dinosaur had walked up the street. Since then, several more living Coelacanths have been caught, all of them exactly as they were when the last Coelacanth fossil was laid down 70 million mythical years ago.

Australian Lungfish

Coelacanth

The stability of today's living creatures which appear in the fossil record was brought up-to-date in a splendid article by Marvin L. Lubenow, M.S., Th.M., presented as a paper at a meeting of the American Scientific Affiliation, Stanford University, in August, 1979. It was printed in the *Creation Research Society Quarterly* of December, 1980, titled "Significant Fossil Discoveries Since 1958: Creationism Confirmed."

Among many things, the article tells us that recent finds of pre-Cambrian micro-fossils (such as blue-green algae supposedly 900 million years old) are virtually identical to present-day organisms.

The article also informs that, from 1976 to 1978, fossils of jawless fish were found in Cambrian rocks, and comments:

> For Creationists, this discovery of fishes (vertebrates) in the Cambrian is, without question, the most significant fossil discovery in the period 1958-1979. The evidence is now complete that all of the major categories of animal and plant life are found in the Cambrian.

Lubenow has much to say about discoveries of fossil land plants. Evolutionists require about 150 million years for marine plants to give rise to land plants. Lubenow says:

> Few people, other than botanists, appreciate the radical difference between marine and land plants and the tremendous changes that had to take place before plants could survive on the land, assuming they evolved.

For instance, aquatic plants are supported and nourished by the water; but land plants have to develop deep roots for support and also for extracting nourishment from soil and transporting it above ground. The land plants would also have to develop immediately a vascular system to conduct the nourishment up the stem and to the leaves, something not needed by aquatic plants which absorb water and minerals directly from the surrounding water. Land plants would have to develop instantly the xylem, the tissue which gives support and rigidity, and which also serves as the vascular system.

If a marine plant became a land plant, it would immediately need a cuticle, a wax-like layer which prevents a land plant from drying out, but which would be a nuisance to a marine plant.

All these changes and adaptations would have to happen immediately when the plant changed from water to land. The immediacy is just glossed over when evolutionists say that land plants evolved from marine plants. But evolutionists do designate immense time for the evolving, and ignore evidence for land plants before the Silurian age.

Lubenow discusses various discoveries of vascular (land) plant fossils so early in the evolution time scale as to make any idea of evolution of plants untenable. His paper "has documented thirty-two individual localities where discoveries of land plant fossils have been made in the Cambrian or below" (i.e., pre-Cambrian). Included were flowering plants (angiosperms), which, being the most complex, were supposed to be the last plants to evolve. He quotes reports by Ghosh and Bose of finds of angiosperms in Cambrian rocks in Punjab, India and also in Kashmir. He quotes Clifford Burdick's reports of angiosperm pollen grains in pre-Cambrian shale of the Grand Canyon.

Lubenow's fossils are unkind to evolutionists—down in the Cambrian "beginning" (and before) flowering trees and conifers appear where only "primitive" algae should be; and jawless fish are there, 100 million years before any fish "evolved." What can an evolutionist do but wish they would go away? Or hope that this does not become generally known!

To sum up the living and the dead evidence: Living creatures are living evidence against evolution, while extinct creatures present gaps between kinds which rule out evolution. We have noted that these gaps forced Professor Goldschmidt to propose his fantastic hypothesis of "the hopeful monster" and that the leaders of evolution are being forced by the fossils to fall into line behind Goldschmidt. The bankruptcy of evolutionism is becoming apparent.

In 1978, one of the world's leading geneticists, Professor Jerome Lejeune, was in Australia. He stated that genetic science has put neo-Darwinism into the museum of obsolete ideas. He said that knowledge of genetics has shown that there could not possibly have been a process of gradual gorillization of something pre-gorilla into something gorilla; nor of something pre-human into something human. He said that the making of any new species would require that, at a definite point in time, from a common ancestor, a distinct new species appeared, which lasted for perhaps thousands of generations. Then, at another particular point of time, by a distinct change of chromosomes, another new species appeared; and so on.

Professor Lejeune said that the making of a new species requires a very precise change of the chromosomes, which could not happen through a group. It has to happen on a very tiny stem, and that means the smallest number of first parents. For a plant species, it means one individual plant; for an animal species, one couple. He acknowledged that this fits the concept of Adam and Eve.

Professor Lejeune's statement helps our case; but it also poses a challenge. It helps our case by quashing both neo-Darwinism and Polygenism, and thereby vindicating the Biblical teaching that humanity started with one man and one woman. The challenge is that Lejeune seems to go close to Goldschmidt's "hopeful monster" proposal, or evolution by big jumps.

I think it is not difficult to answer the challenge. When

Professor Lejeune suggested new species by big genetic jumps, he was obviously stepping outside the strict discipline of genetic science. He was hypothesizing, perhaps trying to collate the facts of genetics with the non-facts of the evolutionism which dominates our halls of learning. It is noteworthy that, at the Chicago convention of 1980 (already discussed), there was resistance from geneticists to "hopeful monsters" and evolution by big jumps.

As an impeccable genetic scientist, I think Professor Lejeune would endorse the views expressed by Sir Peter Medawar that the hard facts of genetic knowledge are that there is NO genetic process known to science that could produce the genetic improvement required for upward evolution. I think both these eminent scientists would certify that the notion that new species arose by macromutations is contrary to scientific knowledge. Which all means that the first human couple could not have risen by genetic process from non-human progenitors. The same rule applies to the first gorillas, cockroaches, algae and the rest.

The Horse

The horse series has been claimed to be irresistible evidence for evolution. There have been various horse series. Douglas Dewar stated that he had seen 20 pedigrees of the horse, and every one was different.

Our diagram shows the series proposed by Professor Gaylord Simpson, generally regarded as an authority in this field. Like all modern horse series, it starts with Eohippus or Hyracotherium (we can regard the two as equivalent).

Hyracotherium, or Eohippus, was about the size of a fox. It was originally classified as related to Conies (Hyrax), and pigs and rodents. It had four toes on its front feet and three toes on hind feet, just as the modern, living hyrax have. Later on, when Darwinism had become popular, it was claimed to be in the family of the horse, and became the

A HORSE SERIES

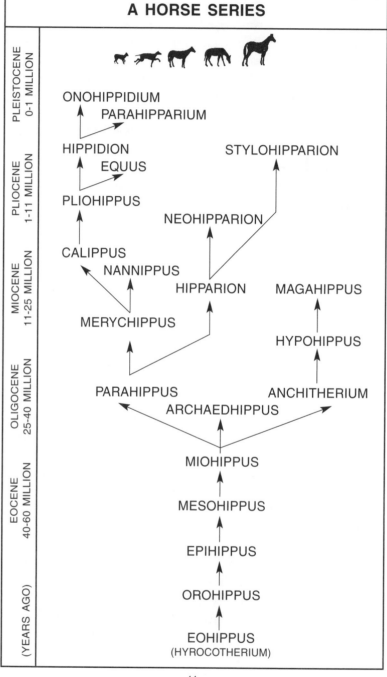

PLEISTOCENE 0-1 MILLION

ONOHIPPIDIUM
PARAHIPPARIUM

HIPPIDION
STYLOHIPPARION

PLIOCENE 1-11 MILLION

EQUUS

PLIOHIPPUS

NEOHIPPARION

CALIPPUS

MIOCENE 11-25 MILLION

NANNIPPUS

HIPPARION
MAGAHIPPUS

MERYCHIPPUS

HYPOHIPPUS

OLIGOCENE 25-40 MILLION

PARAHIPPUS
ANCHITHERIUM

ARCHAEDHIPPUS

MIOHIPPUS

EOCENE 40-60 MILLION

MESOHIPPUS

EPIHIPPUS

OROHIPPUS

(YEARS AGO)

EOHIPPUS
(HYROCOTHERIUM)

start of the horse series. There is no known ancestor of Hyracotherium, or Eohippus; and there is no evidence that it produced any horsey descendants.

The same applies to the other stages of the series. There is no real evidence that any specimen is related to the next specimen. Each link is isolated. These supposed horses are extinct mammals found in different parts of the world, and probably of different species and genera. They have been set up in arbitrary sequences and called evolving horses. Different authorities have set them up in different and conflicting sequences.

Dr. G.A. Kerkut of the University of Southampton (Dept. of Physiology and Biochemistry) is of the evolutionist school. But his book, *Implications of Evolution*, exhorts evolutionists to re-examine several fundamental assumptions on which evolution is based. He seriously questions the horse exhibits. He also questions the validity of the various genera and "the relative positions of these animals to one another, together with some indication as to the accuracy of the relative dating."

Dr. Kerkut also comments that, at present, "*it is a matter of faith that the textbook pictures are true.*" This means that the artists have drawn all those pictures of evolving horses, not from the evidence, but from evolutionist faith.

Eohippus, or Hyracotherium, is a shaky starting point. Dr. Kerkut says that it is not clear that Hyracotherium was the ancestral horse. Several evolutionists, including Professor Simpson, have admitted that Eohippus (Hyracotherium) could just as easily have been the ancestor of the tapir or the rhinoceros instead of the horse.

The late Professor H. Nilsson, a world figure in genetics, in a criticism of the horse series, indicated that this supposed ancestor of all horses, this "dawn horse," is alive today in the Middle East and Africa. Of course, he does not look like a horse—and you must not call him a horse, but the little fellow is there. He is the Hyrax, about the size of a fox and with many toes, darting in and out of thickets. What this means is that Eohippus (Hyracotherium) is alive today, and is not a horse at all. He is the lively little quadruped, Hyrax.

Nilsson points out that the specimens did not become successively bigger. Orohippus and Epihippus are smaller than Eohippus. Then there is an abrupt jump up to Mesohippus, about the size of a sheep and who has lost one of the four front toes. With Mesohippus there is a difference from the previous "horses." A new type started lasting 20-30 million years (evolutionists' years), leading to Parahippus. Then this type disappeared. There is a complete break, an evolutionary gap; and then a very different type, a horsey type, abruptly appears with Merychippus and Hipparion. One-toedness dominated, though clear rudiments of two side-toes may occur. Suddenly there are real horsey teeth. These are such different animals they break any evolutionary sequence with everything preceding them. "Here one cannot speak of evolution," says Professor Nilsson, because there has been "the complete extinction of an ungulate fauna and the sudden appearance of another—and this one richly differentiated . . ."

The reality is different from the textbooks. The horse series is arbitrary; it is artificial. There is no generic connection between specimens. And, something we must specially note is this: the horse series is composed of three distinct sections, of which only the third section includes horses.

The horse series started with three genera. It has grown to 26 genera. Commenting on Professor Nilsson's paper, Frank W. Cousins, engineer and astronomer, writes, ". . . no one yet knows how to place the alleged 26 genera in relation to themselves . . ." (*Speak to the Earth*, p. 102).

So, different authorities set up different family trees. Even Equus is given a choice of immediate ancestors.

Professor Simpson, in *The Meaning of Evolution*, says:

Although there is, in fact, no strictly straight line

of horse evolution, the nearest thing to one leads not from Eohippus to Equus, but from Eohippus to an extinct group decidedly different from living horses and called Hypohippus.

So, all this fuss and "evolving" does not, after all, produce the modern horse, but something quite different called Hypohippus; 3-toed, un-horsey and extinct. It all starts to look like a ponderous exercise in solemn horseplay.

Reptile into Bird

If evolution happened, then some reptiles became birds, and there should be plenty of fossils of intermediates, part-reptile and part-bird, recording the tremendous transformation. It would involve changing the entire skeleton; a completely different nervous system, digestive system and elimination system. It would need a radically different muscular system with powerful flying muscles attached to a large keel bone. Front legs would change into wings. Reptile scales would change into wonderful feathers. There is not a scrap of fossil evidence for any of this. Evolutionists accept it with a pure faith which leaves religious faith for dead.

For a long time evolutionists have claimed that there is a transitional fossil bird, a strange bird called *Archaeopteryx* which is extinct. Some claim that it was a feathered reptile. Recent finds have removed Archaeopteryx from the argument, but, as he is still being used by some evolutionists, such claims must be precisely rebutted.

Archaeopteryx had teeth, but some other extinct birds had teeth. He had a long tail of about 20 vertebrae each bearing a pair of feathers, but, as Douglas Dewar (ornithologist) points out, a long tail does not specify a reptile. He had small claws on his wing tips; but there are two living birds which, in their young stage, have claws on their wings; the hoatzin of South America, and the touraco of Africa. (See

Evolution—The Fossils Say No! by Dr. Duane Gish, 1972, p. 62).

It is claimed that the head is reptilian. In rebuttal, Douglas Dewar stated: "I deem the skull typically avian, and I believe this opinion is shared by the vast majority of evolutionists who have gone into the matter." He quotes Sir Arthur Smith Woodward, an eminent evolutionist, who wrote: "The skull is shaped like that of a typical bird." (See *The Transformist Illusion* by D. Dewar, p. 51).

Sir Gavin de Beer listed the papers of scientists on Archaeopteryx. Six called him a lizard, eight said intermediate, but 37 classified him as a bird.

Reptiles are cold-blooded, birds are warm-blooded; and Archaeopteryx was warm-blooded. His fossils showed the clear imprint of feathers. Archaeopteryx had feathers, perfect feathers, and that clinches that he was a bird.

Archaeopteryx

Things like horns and hooves and hair happen to many kinds; but feathers happen only to birds. Only birds have feathers. Feathers are a problem for evolution; they are miracles of natural engineering. There is a main shaft with dozens of barbs coming from it at a slant. A microscope reveals that the barbs carry hundreds of thousands of little barbules. And, on the barbules, are millions of tiny barbacelles. Then, on many of the barbacelles, are the tiniest hooks which, in flight, interlock the whole wing into a plane of wonderful firmness and elasticity and lightness. Then, in an instant, the whole structure can be suddenly unlocked to let the air pass through.

So, how did a reptile get wonderful feathers? Some will claim that the reptile scales *mutated* into feathers. On good authority, that would not be a mutation, it would be a miracle. Some claim that friction of air did it; friction of air on the scales of generations of reptiles as they jumped from tree to ground, or reptiles which ran along the ground waving their forelimbs. Friction of air, over vast time, fraying scales into wonderful feathers: what do you think?

Douglas Dewar, Fellow of the Zoological Society, said: "If this is serious science, then so is the story of Cinderella."

As mentioned, recent finds have ended the argument about Archaeopteryx. *Science News* (Sept. 14, 1977) reported the finding of a true bird older than Archaeopteryx by foremost palaeontologist, James A. Jensen. However, Jensen has since decided that the bones are of a flying reptile. The latest position is set out in the previously mentioned article by Marvin L. Lubenow (*C.R.S. Quarterly*, December, 1980). Lubenow reports that Jensen has since found other bones which are bird bones. It seems that "Jensen feels that he needs much more material before he can report these discoveries with any specifics."

Lubenow's article says:

> Although Archaeopteryx continues to be the oldest fossil bird discovery . . . it must now share antiquity with another more modern type bird fossil discovery . . . The discovery of a more modern type of bird that was contemporary with Archaeopteryx seems both to indicate a greater antiquity for birds, and to effectively

remove Archaeopteryx from the role of a transitional form.

My interpretation is that, regardless of whether the first find was a bird or a flying reptile, the second find looks like a true bird contemporaneous with Archaeopteryx; and that removes Archaeopteryx from the controversy. (Also see *Appendix D*, pp. 183-184).

Chapter 4

"HUMAN EVOLUTION"

We come to the ape-men, and this is the point at which evolution penetrates religion—at the origin of man.

Evolution leads logically to polygenism, that is, not one first man Adam, but groups of brutish men and women mutating from parents who were not human.

Polygenism plays havoc with the central Christian dogma of Original Sin. And lately under the influence of Teilhard de Chardin's evolutionary theology, all the Catholic dogmas are being turned upside-down. If we can demonstrate that the ape-men never existed, then the whole case for evolution and for polygenism and for Teilhardianism collapses.

Adam

Ape-Man

Ape-Men: The average person now believes that there were strange creatures in prehistoric times not quite men and not wholly animal. We are being told, in the name of science, that these ape-men existed and that we come from them.

If this theory is wrong, then our era is really the darkest age of all. But right or wrong, the theory is successful. It is so successful, that Adam and Eve are laughed out of court. We are witnessing the triumph of a very deep game, and the aim of the game is to get rid of God and undermine the Bible.

Adam was created in the image of God, physically perfect and, before his fall, intellectually sublime. The ages of faith produced the Christian images of Adam and Eve in devout profusion on the walls and windows of the great cathedrals and the ceilings of churches and chapels.

But today the picture is changed. The brute-man is today's concept of Adam, or many Adams.

This brutish Adam has changed the whole world outlook and philosophy. That is the extraordinary achievement of Darwin. Whether Darwin was right or wrong, he has changed man's concept of man.

There have been various family trees of man, disagreeing and conflicting with one another.

At the base of just about every family tree of man was Ramapithecus. However, because of several dramatic discoveries since 1972, we'll leave Ramapithecus until later.

Let's start in the 1920's with the triple blunder: Nebraska Man, Piltdown Man, and Neanderthal Man. In 1925 humanity's very existence depended on these three gentlemen; the first American, the first Englishman, and the first European. Let's see what became of them.

Nebraska Man: In 1922, someone found an unusual molar tooth in Nebraska. The eminent Professor Henry Osborn, then of the Natural History Museum in New York, firmly declared that this tooth must have belonged to a creature half-ape and half-man. Many specialists and scientists agreed. And so there he was. And America was proud to have a 100% all-American ape-man.

Nebraska Man

Nebraska Man lasted a total of five years. In 1927, further discoveries proved that the unusual tooth on which Nebraska Man had been built had come not from an ape-man, but from a peccary, a sort of wild pig. Professor Osborn's ape-man was a pig! Such is the fantasy world of evolution.

Neanderthal Man: He was supposed to be little better than the chimpanzee: small brain capacity, walking stooped forward with knees bent, hairy, animalistic and very disagreeable.

The mistake about brain capacity was corrected by the great Marcellin Boule. Boule proved by measurements that

Neanderthal Man had a bigger brain capacity than modern man. Then, evidence was found that he believed in the supernatural. Douglas Dewar tells us that Neanderthal Man buried his dead with ceremony and that he was a skilled craftsman.

Also, there was evidence indicating that Neanderthal Man had intermarried with modern-type man. And Fontechevade fossils proved Neanderthal Man and modern-type man were contemporary.

To cap it all, in 1929 Professor Sergi proved that Neanderthal Man walked perfectly upright—as upright as any of us today. He could have stood at attention for any "prehistoric" sergeant. Pictures and statues representing Neanderthal as a shambling brute are the artist's own work. The artist can do anything. He can make Neanderthal look like a brute or a philosopher; it all depends upon the artist.

Neanderthal man was a hunter and a nomad who used caves. But you can be sure that he was just another race of homo sapiens, your brother, a real man with a soul to save.

Piltdown Man of England: For 40 years he fooled the world and helped to scuttle Adam and Eve.

The principal characters in this drama were Charles Dawson, who found the first part of the skull, Sir Arthur Smith Woodward of the British Museum, and a student priest, Father Teilhard de Chardin. In December of 1912, Dawson and Woodward told a distinguished audience that during four years of work, they had found strange fossils at Piltdown, namely, the upper part of a skull which was human, and nearby, a broken lower jawbone which was quite ape-like except that the teeth had worn down the way human teeth wear down, not as an ape's teeth would wear. A tooth was missing, an important canine tooth was missing. If that missing canine tooth could be found, and if it were worn down like the molars, then the ape-man's case would be strengthened.

The tooth was found eight months later, August 29, 1913. Teilhard de Chardin had returned from France. Next day,

Dawson, Woodward and Teilhard went down to the Piltdown site to sieve the gravel. After a time, Teilhard called out that he had found the missing tooth. The tooth fitted the jaw and Mr. Piltdown was established. Some scientists, however, could not accept that the ape's jaw belonged with the human skull, and these included leading anthropologist Marcellin Boule.

However, in 1915, Dawson reported that, in another field two miles from Piltdown, he had found two small pieces of skull and a molar tooth similar to the Piltdown fossils. That did it. An apelike jaw and a human cranium at Piltdown, and a similar find at a site two miles distant. This went beyond coincidence. The ape-man must be genuine. Even Boule softened a little. Piltdown man was established: the earliest Englishman, half-a-million years old.

Piltdown Man

He reigned for 40 years. Then some skeptics insisted that he be retested. He survived the first test. But his age was dropped from 500,000 years to 50,000 years. The critics forced further tests. And now it was a disgrace. The skull was that of a modern man, a completely modern man. The jawbone was from an ape that had only recently died. Now, too, they saw that the teeth had been filed to make them look human. The marks of an abrasive were discovered. Now they saw that the jawbone and the teeth had all been stained by chemicals to make them look like ancient specimens. Now they saw the crudeness of the fake. Why had they not seen it before? And that skull, dated by experts first at 500,000 years and then at 50,000 years, then at a few thousand years—now it is 500 years old.

Who perpetrated the fraud? Woodward seems victim rather than culprit. Dawson, the non-scientist, was the obvious target; and the official verdict pinned the guilt on Dawson (already deceased). But that did not stop the speculation. In 1978 it was announced that a tape recording by Professor James Douglas (who had recently died at the age of 93) states that Professor William Sollas was the hoaxer. However, Dr. Halstead, who has the tape recording, has written to *The Times* and has stated that several of the British Museum authorities of the time were involved. In a radio interview Dr. Halstead mentioned that it is fairly certain that the fake jaw came from some unregistered orangutan bones in the British Museum.

Since the bubble burst, Piltdown has been a prolific source of "who-dunnit" writings, and the weight of them is bearing down on Teilhard de Chardin as the mastermind in the hoax. Two of the latest writings are representative.

In a recent article, "The Piltdown Conspiracy," in *Natural History Magazine,* prominent evolutionist Stephen Jay Gould writes: "I have also remarked, both with amusement and wonder, that very few believed the official tale of Dawson acting alone . . . several of the men I most admire suspected Teilhard . . ."

Then Gould proceeds to sift the clues and makes a strong case that Teilhard acted with Dawson, possibly with some young subordinates of the British Museum involved. Dawson died and events happened which caused Teilhard to remain silent instead of revealing the joke. Jobs and careers would have been harmed, including Teilhard's own unexpected new career. Gould asks:

> Shall we then blame Teilhard or shall we forgive him? We cannot simply laugh and forget. Piltdown absorbed the professional attention of many fine scientists. It led millions of people astray for forty years. It cast a false light on the basic process of human evolution. Careers are too short and time too precious to view such waste with equanimity.

Gould charitably excuses Teilhard on the grounds that he acted as a joker, free of malice and not seeking reward. But the joke went sour. Gould thinks Teilhard "suffered for Piltdown throughout his life." He thinks Teilhard "must have cried inwardly as he watched Smith Woodward, and even Boule himself make fools of themselves—the very men who had befriended him and taught him."

Piltdown is treated in books by evolutionists and by anti-evolutionists. Among the latter, Malcolm Bowden's *Ape-men: Fact or Fallacy?* (1977) devotes a large section to probing the Piltdown case, with threads leading to Teilhard. Bowden makes the predication: "It is with considerable hesitation that I state the case against a man who has achieved such worldwide fame and is venerated by many"; but he makes his

> Final Conclusion: As it is over sixty years since the Piltdown excavations took place, it would be extremely difficult to say with absolute certainty the true identity of the hoaxer. It is submitted, however, that until further facts become available, the evidence given in this section points to the instigator of the fraud being Pierre Teilhard de Chardin, S.J.

Later in his book, Bowden has something to say about

Teilhard's implication in the Peking Man affair in China and in other evolution operations.

Tracking down the mastermind of Piltdown is an interesting exercise, but I suggest that many leading scientists of the day were also culprits. They allowed a detectable forgery to be foisted on the world, spreading the creed of evolution for 40 years.

The Australopithecines: Now we move on to the heavy-weights from Africa, from Java and from China. From Africa come the Australopithecines. In 1924, Dr. Raymond Dart found his famous Taung skull in a rock quarry of South Africa. Later on, Robert Broom and J. T. Robinson found other fragments which they regarded as the remains of hominids.

Australopithecines are pictured as large-jawed, small-brained, standing about four feet tall and walking in approximately human fashion, not yet men but a pre-human phase of hominid evolution. To the public it all sounds scientifically definite. The public would be surprised to know how little fossil evidence there really is—and surprised to know how scientists sometimes allocate the bones the way they have preconceived them to be, according to the theory of evolution.

In 1954 Sir Solly Zuckerman spoke out. He had made a detailed study, bone by bone, and he gave his verdict on the basis of brain-capacity, jaws, teeth and the point of balance of the skull on the spine. Lord Zuckerman's verdict was that these animal bones gave no evidence of something evolving into a human. In scientific stature, Lord Zuckerman would overshadow the others and he evidenced no bias. He has had a brilliant career leading to his appointment as Chief Scientific Advisor to Her Majesty's Government, and then, in 1971, to his elevation to the Peerage as Lord Zuckerman. His verdict should have carried great weight.

His verdict was fatal to Australopithecus, so it was

scarcely heard. In 1974, Dr. Charles Oxnard of Chicago University performed a multivariate computer analysis of Australopithecine bones. In 1975, he announced that "Australopithecus was uniquely different from both man and modern ape. If anything, it tended towards the orangutan." As we shall see, discoveries since 1972 have pushed the Australopithecines into a dead end. They are now generally regarded as not ancestral to man.

Let us look briefly at the other still-prominent members of our supposed family tree.

Dr. Louis Leakey worked in Africa. In 1950, his wife found 400 fragments of a skull. Actually, there is sound basis for believing that parts of two skulls were involved. But they put the 400 pieces together into a reconstructed skull. The jaw was missing, so they added a modeled jaw bone based on a jaw that Dr. Leakey's son had found in another place. The result was the famous Zinjanthropus, or Zinj. The popular magazines published portraits of Zinj and stories of his personal habits.

Zinj was declared human, the oldest "real man," 1,750,000 years old. The excitement was so great that it was stated that, when Leakey found Zinj, the textbooks became obsolete overnight.

But Zinj soon fell from glory. Everyone, even the Leakeys, agreed that Zinj was not human after all. He was just another Australopithecine, just a brute.

Dr. Leakey produced another candidate, Homo Habilis (Clever Man), clever with tools. From pieces of skull, parts of jaws, teeth, some foot bones and some finger bones, Leakey considered that he had found a new genus Homo, with skull capacity about 670 c.c., who was a tool user because some simple stone tools were found nearby.

Many evolutionist authorities strongly disagreed with Dr. Leakey's classification of Homo. On the scanty evidence they claimed that Habilis was only a variant of the Australopithecines. Within the disputing schools of thought, the idea began to appeal of an evolutionary sequence of Australopithecus to Homo Habilis to Homo Erectus (Java Man and Peking Man) and then to Homo Sapiens. This neat sequence was upset by Dr. Leakey himself when he reported that he had found the remains of Australopithecus and Habilis and Erectus, all contemporaneous, in Bed II of Olduvai Gorge.

Java Man: We move on to Homo Erectus, a classification which includes Java Man and Peking Man. Firstly we look at Java Man. In 1891, Dr. Eugene Dubois of Holland gave up his career and went to Java to search for the missing link. Later he presented to the world his Pithecanthropus Erectus, or Java Man. Java became a hero, talked about in the same way as were Pitt and Napoleon. Popular histories published detailed portraits of him. G. K. Chesterton commented that no uninformed person looking at his carefully lined face would imagine that this was the portrait of a thigh-bone, a few teeth, and a fragment of cranium.

Pithecanthropus Erectus   *Neanderthal Man*   *Cro-Magnon Man*
    (Java ape-man)

How did it happen? In 1895 when Dubois returned to Europe, he showed to an International Congress of zoologists what he had found in a river bed in Java: a skullcap and a tooth which both appeared to belong to an ape. He also showed them something that had been found a year later and about 50 feet distant, namely, a thigh bone that seemed to be human.

Now, if you or I found an ape's skullcap in one place, and, a year later, we found a human leg bone in another place 50 feet distant, I do not think we would get excited and say they belonged together: "Look! We've found an ape-man!" However, Dubois said they belonged together and scientists let him say so, because it was believed that man had only recently migrated to Java. So, assuming there had been no humans in Java, they allowed the ape's skullcap to belong to the human thigh bone, and there was the "missing link" which they wanted to find.

However, Dr. Dubois had not told the whole truth. He had not told the most important part of the story. He did not tell that he had also found two human skulls in the same stratum as the skullcap. To have told this would have spoiled his case because those human skulls, the Wadjak skulls, as they are called, showed that real human beings did live in Java at the same time as the supposed ape-men. And that would have meant that there was no need to link the thigh bone with the skullcap 50 feet distant. And that would have meant that the evidence of the alleged ape-man vanished. For something like 30 years, Dubois kept the human skulls secret and hidden. This was inexcusable. It caused the great biologist W. R. Thompson to say that the success of Darwinism was accompanied by a decline in scientific integrity.

About 1921 certain events persuaded Dubois to reveal the human (Wadjak) skulls. But, by then it was too late. Java Man was established and had done his damage. But you should be told that the leading authority on fossils, Marcellin Boule, rejected Java Man. He said it was a gibbon. Even more significantly, Dr. Dubois himself renounced his own Java Man in 1938.

In addition to the exploits of Dubois, there were two later expeditions to Java. The Selenka expedition, 1907-08, was conducted with strict scientific discipline. It excavated more than 10,000 cubic meters of earth, to a depth of 12 meters at the site of the Dubois finds. It collected 43 large boxes of fossil bones. It reported that volcanoes on the island would quickly fossilize bones, which meant that fossils were not necessarily very old. It remarked that violent floods would confuse the geologic strata. It found evidence indicating human existence in the same stratum as the supposed ape-man. It found *no* evidence supporting Dubois' ape-man.

The Selenka findings were a setback to the "missing link." Then around 1921, Dubois revealed that he had kept hidden the vital Wadjak skulls—human skulls—which nullified the linking of the skullcap with the thigh bone. The ape-man was unstuck unless something was done.

So began the third expedition. In 1931, G. H. R. von Koenigswald was sent to search the area. He found a number of human skulls, but no "missing links." Because of the financial Depression, von Koenigswald lost his job with Geologic Survey. In 1936, through Teilhard de Chardin's influence, he was made a research associate by the Carnegie Foundation in America and was granted considerable funds to search for fossil man.

Malcolm Bowden's book (p. 140) tells us:

> Teilhard's international contacts were growing and became so extensive that Cuenot says: "One has the impression of a vast web, of which Teilhard held in parts the threads, where he served as liaison agent, or better still, as chief of staff, able, like a magician, to make American money flow, or at least to channel it for the greatest good of palaeontology."

Thus von Koenigswald was returned to the quest. By 1938

he had found fragments of jawbones, some teeth, fragments of skulls and a skullcap. From these he produced Pithecanthropus II, III, and IV. What was their true scientific worth? Dr. Duane Gish (in his book *Evolution: The Fossils Say No*) refers us to Boule and Vallois (in their *Fossil Men,* 1957, pp. 118-122). They judged that these skulls had the same general characteristics as that of which Dubois had made his Java Man; and they regarded the Java Man skullcap as "very similar to those of chimpanzees and gibbons."

By this time, Dubois was repudiating his own Java Man. He declared that, after long study, he was of the opinion that "we are here concerned with a gigantic gibbon." Ironically, now that he was trying to make amends, he was dismissed as unreliable. He was not allowed to kill off his own Java Man, which was now enshrined (together with Peking Man) as "Homo Erectus," the beginning of man.

"Ape-men" are hatched in the shadows, shielded from the light of truth. Official publications veil the scandals. Textbooks omit the real evidence. They are still solemnly teaching that the ape's skullcap and the human thigh bone belong together. Historian Francis Vere lamented, "We are entitled to *all* the evidence . . . If the evidence is kept from you, how are you to find the explanation?"

Peking Man: The obsession to find ape-men has truly been "accompanied by a decline in scientific integrity." The affair of Peking Man is long and complex. We must condense and simplify it.

Dr. Davidson Black, in 1914, had helped put together the Piltdown skull. In 1926 while the Piltdown "ape-man" still reigned in England, Black was in China and was Professor of Anatomy in the Medical College in Peking. Also, he represented the Rockefeller Foundation of America, and from it he obtained an annual grant of $20,000 (a lot of money in those years) to continue explorations which had begun at Choukoutien.

Dr. Black took charge of the excavations. A Chinese scientist, Dr. Pei, was in charge of field work. Father Teilhard de Chardin, of Piltdown renown, had come under Church censure for his extreme evolution philosophies and had gone to China. He was put on Black's team as adviser.

In 1927 a molar tooth was found. Black pronounced that it was part ape, part human, and a new ape-man was launched, Sinanthropus pekinensis (Peking Man). Black rightly judged that the waiting world wanted ape-men and would not be too critical. The press welcomed a new ancestor based on one tooth.

Next, in 1929, something of a skull was found. Black hailed it as justifying what he had predicted from the molar tooth. He had confirmed his Sinanthropus (Peking Man). But what did he *really* have?

Telihard immediately reported to France the finding of a skull which "manifestly resembles the great apes closely," and with small brain capacity. This assessment was corroborated by three other experts who later examined the skull. It should have been left at that; the skull, not of an ape (for no fossil apes have been found in China), but of a baboon or a large monkey, skeletons of which had been found in that area, similar to, but larger than modern living specimens.

Dr. Black, however, wanted Sinanthropus. He made a model, not a cast but a model of the skull, and he wrote a lengthy document describing it. According to Father Patrick O'Connell who was in China and who made a special study of the Peking Man affair, Black's model "is an artificial model of the skull of the mythical Sinanthropus, not a cast of the skull described by Father Teilhard de Chardin (and others)." The model, and the "equally artificial" document purporting to describe the model, had the aim of representing Peking Man as an intermediate between Java Man and Neanderthal Man.

Black made the brain case 960 c.c. (on his calculations), which is in the human range, and which is far in excess of the skull described by Teilhard. From accounts by Father O'Connell and Malcolm Bowden, Black's methods in the reconstruction of this head of Sinanthropus were heavily tainted with that lack of scientific integrity. In other words, the model was false; Black made it represent what he wanted it to represent.

As they continued to excavate they uncovered two great heaps of ashes, and in the ashes were bones of numerous animals. Also in the ashes were some more broken pieces of monkey-like skulls which Black claimed were more of his Peking Man. A remarkable aspect was that parts of monkey-like skulls were found, but scarcely any of their skeletal bones. This indicated that only the heads of the monkey-like creatures got into the "cave." That indicated that the heads had been carried in. There was no fossil evidence of this creature's posture or stance. Yet, Teilhard confidently pronounced that this hominid walked upright and was two-handed.

The world was told that "traces of fire" had been found. The picture of Peking Man emerged as a creature just across the dividing line; a human, but only just a human; using stone tools, walking upright, living in a cave, and using fire for cooking. Evidence was twisted and suppressed to fabricate Peking Man; the press cooperated, and the world fell for it.

An eminent palaeontologist, Professor Breuil, visited the site in 1931 at Teilhard's invitation. When Breuil returned to France, he published an article which revealed that the "traces of fire" were actually the remains of industrial furnaces. Also, he had seen the fossils themselves, and he raised the question (which other experts would ask later): Could creatures with such animal skulls have worked such a big industry?

Dr. Black and Teilhard de Chardin must have known what Breuil had published in 1932, and Breuil was a world leader in archaeology and anthropology. Yet, they published their own book, *Fossil Man in China,* the following year, which claimed to list all papers published on the subject up to 1933, but which left out Breuil's important publication. Referring to this omission, the historian Francis Vere wrote: "This is simply incredible. One can only conclude that Breuil's discoveries, being inconvenient, were deliberately suppressed." (*Lessons of Piltdown,* p. 47).

What they were calling "traces of fire" were really two enormous heaps of ashes. One heap on a lower level was not fully uncovered. The heap on the upper level was an ash heap as long as a football field, half the width of a football field, and, even after centuries of compression, the height of a two-story building. These were the remains of industrial furnaces which had been fired continuously for long periods. Also, there were thousands of quartz stones brought to the site from a least a mile away. The evidence was clear that advanced humans had worked an industry of limestone burning, possibly in the building of the ancient city of Cambaluc near where Peking now stands. But Black and Teilhard persisted in playing down the extent of the fires, presenting them as small hearth fires and holding back information on the existence of industry.

Peking Man was represented as living in a cave. There was no cave. On the hillside there were two levels from which limestone had been extracted. An old landslide had apparently covered everything, and now they were calling them caves.

The eminent Marcellin Boule was not averse to evolution in theory; but, as we have seen, he was always critical of methods that lacked scientific integrity. Boule was invited to visit the site of Peking Man, and he did visit it. The great man's reaction was one of annoyance at having his time wasted with monkey skulls. Boule's firm opinion was that real men had worked a large industry there. Boule rejected Dr. Black's theory outright. He called it a fantastic hypothesis—the idea that the monkey-skulled creatures could have operated such an industry. Boule declared that the monkey-like skulls and the other animal bones mixed up in the ashes were the remains of food eaten by human workmen, who had then tossed the skulls and the bones into the ashes.

The opinions of Boule and Breuil should have ended this latest escapade of prospecting for ape-men. Instead, the tactic of twisting and suppressing the vital evidence persisted, while the voices of Boule and Breuil were scarcely heard. The world was systematically deceived, and Peking Man grew stronger on the diet of deceit. So also did Dr. Black's world stature. By 1934, he had received the honor of Fellow of the Royal Society of London. Imagine the reaction if real human remains were discovered on the "Peking Man" site. Well, that is exactly what had happened just before this.

Toward the end of the 1933 season, three human skulls were unearthed in the "upper cave." They were delivered to Black's laboratory. Here were human remains, not battered fragments like Sinanthropus, but three complete adult skulls with jawbones; also other bones totalling probably six persons, including a child. This significant find justified the views of Boule and Breuil, but boded ill for Sinanthropus.

Teilhard sent a report to France that human bones had been found; but, without going to investigate the site, he added that these real men must have been of later date, with no bearing on Sinanthropus, and of little interest. Dr. Black had been having heart problems. In March 1934, he was found dead in his laboratory, among the human fossils.

Dr. Franz Weidenreich took charge after Black's death. He proceeded to make his own model of Sinanthropus (Peking Man), and he outdid Black. He used parts of four different skull pieces to build a skull, then had a sculptress mold it into a woman's head, approximately human. The brain capacity was a whopping 1200 c.c.'s. Let us then note that, in 1937, Teilhard (in contradiction of his 1933 report) published an article in *Etudes* which conveyed that no human remains had been found. Instead, he talks of a "great male"

(Sinanthropus) skull of 1200 c.c.'s to refute Boule's view that it needed real men for working the industry. This skull is apparently a model by Weidenreich.

Perhaps you are thinking: Why not examine the actual fossils and get to the truth? We cannot do this, because every fossil bone of Peking Man has disappeared. All that is left are the imaginative models made by Black and Weidenreich. These models have been used by the communists in China to teach the Chinese that they come from apes.

All the fossils have disappeared mysteriously. One explanation is that, *after* the War, the fossils were put on board an American ship, and simply disappeared. Another explanation is that, *during* the War, the invading Japanese destroyed the fossils. This cannot be true. The Japanese allowed Dr. Pei to carry on the work on Peking Man throughout the War. Father Patrick O'Connell has written that the Japanese did not interfere, but that there was good reason to destroy the fossils later. "The skulls were therefore destroyed before the Chinese Government returned to Peking, in order to remove the evidence of fraud on a large scale."

That, to me, was the masterstroke. Get rid of the incriminating evidence, and let Peking Man live on as our immediate ancestor. This has been achieved. Peking Man and Java Man have been bracketed together under the title of Homo Erectus, the first "Man" across the threshold.

Through a "decline in scientific integrity" and propaganda through the mass media and the education system, the world was presented with a neat evolutionary sequence: From Ramapithecus to Australopithecus up to Homo Habilis; then on and up to Homo Erectus; then on to Homo Sapiens (including Homo Neanderthalis). That was until 1972.

Year 1972: Big Changes Begin

Skull 1470: A thunderbolt struck the "tree of man" in 1972. Richard Leakey, son of the late Dr. Leakey, found a skull and some leg bones in Kenya, and everything was changed. The skull was named "Skull 1470." Its brain capacity was about 800 c.c., which is at the extreme low end of Homo Sapiens. Leakey claimed it was human, and he dated it 2.8 million years old. Leg bones in the same stratum were "indistinguishable from those of modern man." In addition, there were tools of man.

Though subject to argument, "Skull 1470" was tentatively classified as genus homo. This was a crisis: a human skull and human bones far older than the supposed ancestors of man; far older than Homo Erectus, Homo Habilis and the Australopithecines. Richard Leakey announced: "What we have discovered simply wipes out everything we have been taught about human evolution; and I have nothing to offer in its place." Since then, he has offered something, which we will come to.

Ramapithecus: The only "ancestry" not affected by "Skull 1470" was Ramapithecus, at the base of the "tree." Until the crisis he was not important; he was too far in the past— 11 million years ago. Then, suddenly, mankind depended on him; he was man's only link with the beasts.

The evidence for Ramapithecus consisted of some fragments of jawbone and some teeth. They had been found in several parts of the world, but were so meager, the whole lot would fit in a cigar box. They were so fragmentary the pieces could be arranged to indicate either hominid or ape, according to one's inclination. His credentials, always flimsy, were invalidated with the discovery of living baboons, in Ethiopia, with jaw and teeth characteristics similar to the Ramapithecus relics. This meant that the Ramapithecus relics were no more entitled to be called hominid than were the living baboons.

The knockout came in 1978, when David Pilbeam found a complete jawbone of Ramapithecus. He wrote that "this new specimen did not conform to our expectations." After describing some of its aspects, he said: "This, together with other data, made it clear that the story of human origins needed rethinking."

With the disqualification of Ramapithecus, "the fossil tree of man" became a blank.

Meanwhile, further discoveries were superseding "Skull 1470." The picture was changing rapidly.

Laetoli (Tanzania) Finds by Mary Leakey: In 1974, Mary Leakey (mother of Richard Leakey) found, at Laetoli, human-like jawbones. She dated them 3-1/2 million years old.

In 1976 she discovered fossilized footprints in petrified volcanic ash which were dated more than 3-1/2 million years old. She reported: " . . . the form of his foot was exactly the same as ours." And again: "Leg structure must have been very similar to our own." Dr. Louise Robbins, a specialist in footprints, analyzed them thus: "Weight-bearing pressure patterns in the prints resemble human ones . . ." Dr. Robbins remarked: "They looked so human, so modern, to be found in tuffs so old." (*National Geographic,* April 1979, pp. 446-456).

A double-page artist's illustration depicts a volcano, and, in the foreground two bestial, naked, hairy creatures, walking (upright, of course) on human legs and feet across a damp ash plain. Behind them they leave a trail of footprints for our fossickers to find eons later. To this, a creationist scientist would comment: "Modern humans make modern human footprints: Your dating methods are useless. The illustration is fantasy."

If jawbones which looked completely human and footprints which were made by a foot "exactly the same as ours" were found in recent rocks, they would be promptly classed as modern man, even by the most ardent evolutionist. However, when evolutionist's radiometric dating makes the rock 3-1/2 million years old, the evolutionists are in a dilemma. They express astonishment and seek sanctuary in evolutionese: "We have found hominid footprints that are remarkably similar to those of modern man. Prints that in my (Mary Leakey's) opinion could only have been left by an ancestor of man." The real evidence is processed to fit the preconcept of evolution. When faced with incongruous evidence, it seems that evolutionists cannot judge it objectively; they are prisoners of their pre-judgment. In contrast, a creationist scientist would say: "Human footprints were surely made by humans; human jawbones surely belonged to humans."

Hadar (Ethiopia) Finds by Johanson: (a) In 1974, Dr. Donald Johanson and Dr. Maurice Taieb found parts of two lower jaws, half an upper jaw and also a complete upper jaw with 16 teeth. He classed them homo (man) and dated them about 4 million years old. (*Scientific American,* December 1974, p. 64).

(b) In 1975, Johanson found the remains of two children and four or five adults, which were called "the first family of early man." There were jaw-bones, teeth, and scores of hand-bones, foot-bones, leg-bones, vertebrae, ribs and partial skulls. (*National Geographic,* December 1976).

The report, with descriptions and photos, convey the certainty that these are human remains. For example, pages 808-809 display photos of some of the finds with comments such as: "hand-bones . . . arranged as a composite pair, bear an uncanny resemblance to our own—in size, shape and function." Also this: "the jaw is U-shaped, like those of humans, not V-shaped like those of Australopithecines." And also this: "for a lark, members of the research expedition made clay casts of their own teeth; one woman's jaw bore a startling resemblance to a three-million year old specimen." There was no doubt in any minds that these were human; and they dated them 3-1/2 million years old.

(c) Soon after these finds, in 1975, Johanson discovered a different creature. He found nearly half the entire skeleton, and dated it 3 million years old. He called it Lucy. He was certain Lucy was beast, not human: "Lucy is far from being a member of the genus Homo. Fully grown, she is less than four feet tall. Her arms in relation to her legs are longer than a modern human's but not as long as an ape's . . . The lower jaw's V-shape and its narrow incisors resemble those of Australopithecus . . ." (*National Geographic,* December 1976, p. 802).

Time reported (November 7, 1977): "They uncovered the fossilized remnants of a 20-year-old female Australopithecus lying in a layer of sediment 3 million years old . . . Lucy was a small creature, not much more than a meter tall, with a brain capacity about a third that of modern man. Lucy's skeleton gave scientists their best clues yet to the proportions of Australopithecus . . . surprisingly short-legged . . . no doubts that she walked erect."

Actually it is doubtful that Lucy really walked erect. Dr. Owen Lovejoy stated in 1979 that a multivariate analysis of Lucy's knee joints showed that they were "distinctly ape and far removed from Man."

There was agreement among the Leakeys, and Johanson and Taieb, and in scientific circles and in the press; there was agreement all round that the above finds (a) and (b) were Homo, and 3-1/2 to 4 million years old; and that (c) was brute (namely, Lucy the Australopithecine) and was later in time than the Homos.

(I use the evolutionists' millions of years in arguing the case, but I do not accept them. In another section the dating methods are critically discussed and rejected.)

Non-evolutionists regard some of the finds at Hadar and at Laetoli as being decidedly human; but, because they were "dated" so old, evolutionists cannot let them be human. That would uproot the "Tree of Man," and they will not permit that.

Erecting a New "Tree of Man"

The Koobi Fora Skull: In 1975, Richard Leakey found

a fairly intact skull at Koobi Fora, near where he had found Skull 1470. He claimed that this latest skull resembled Homo Erectus (Java Man, Peking Man); that it had a brain capacity of 900 c.c., and he dated it 1-1/2 million years old. He stated that this skull was more advanced than Skull 1470. He dropped 1470 from "human" down to Homo Habilis.

Out of all this he proposed his new "Tree of Man," namely, Homo Habilis evolving up to Homo Erectus, which then evolved up to Homo Sapiens.

Skull 1470   *Homo Erectus*   *Homo Sapiens*

Now let us discuss the happenings at Hadar and Laetoli, and Johanson's new "Tree."

Johanson's turnabout and his Afarensis: When Johanson found jaws, teeth and bones at Hadar in 1974-75, including "the first family of early man," and when Mary Leakey found human-like jawbones and footprints at Laetoli, Johanson regarded them all as genus Homo. However, by 1979, influenced by Dr. Tim White, Johanson performed a turnabout. He now classified his Homos, with the brutelike Lucy, as a single creature, a primitive Australopithecus which he named Afarensis, and which would henceforth be the basis of his new "Tree of Man." This tree starts with Afarensis as the "common parent" of two branch lines. He said that one line produced the Australopithecines: first Africanus, then Robustus and on to Boisei (or Zinjanthropus). He said the other branch-line produced Homo Habilis, then up to Homo Erectus and finally up to Homo Sapiens.

The Leakeys reject Afarensis, as do several other scientists. They say it is a composite of at least two separate species.

However, Johanson's Afarensis has been generally accepted as the "parent" of Australopithecines and of Man. But a new find seems to demolish the status of Afarensis and almost everything else.

Another Boisei (or Zinjanthropus): In July, 1986, newspapers reported a new find in Kenya by an associate of Richard Leakey, Dr. Alan Walker. It was another skull of Australopithecus Boisei (or Zinjanthropus); but this one was dated 2-1/2 million years old, which is far older than any previous Boisei specimen. This age is nearly as old as Lucy, yet Lucy was supposed to be the ancestor of Boisei. The new find has sent shock waves through "Hominid Evolution." It means: (1) Boisei could not have evolved from Afarensis (or Lucy); (2) Africanus and Robustus therefore could not have evolved from Afarensis: (3) Afarensis becomes insignificant, and only one of several coexisting species of Australopithecines, and all of them without any known ancestors; (4) Homo Habilis has no known ancestor; and coexisted with the Australopithecines; (5) The slate of "hominid evolution" is wiped fairly clean.

The ramifications of the new find are evaluated in *Discover* (September 1986). The cover shows a cast of the skull and the words: "An extraordinary 2.5 million-year-old skull found in Kenya has overturned all previous notions of the course of early hominid evolution. We no longer know who gave rise to whom—perhaps not even how, or when, we came into being."

The evaluation is by a palaeontologist at the Johns Hopkins School of Medicine, Pat Shipman, who is the wife of Alan Walker, the man who discovered the skull. Shipman's article (pp. 86-93) tells us this: "What the new skull does, in a single stroke, is overturn all previous notions of the course of early hominid evolution." (p. 89). The article tells us that the skull is "older than any of its putative Robustus ancestors. Despite its antiquity, it's already specialized into a Boisei." (p. 90).

Shipman points out that there would not be enough evolution time for Afarensis to give rise to four clearly distinct hominid lineages (p. 91). She raises a big question mark over Afarensis, saying that Afarensis "may be all these different hominid species incorrectly grouped as one. And then where is the ancestral hominid species? The best answer we can give right now is that we no longer have a very clear idea of who gave rise to whom; we only know who didn't. . . . In fact, we don't even know what sort of 'ancestral species' we're looking for." (p. 92).

The article is an admission that evolution scientists have not any evidence about the origin of Man.

Shipman concludes: "Like an earthquake, the new skull has reduced our nicely organized constructs to a rubble of awkward, sharp-edged new hypotheses." Then she adds, rather oddly: "It's a sure sign of scientific progress."

The foregoing deals a fatal blow to the Lucy/Afarensis scenario. Let us now see what is the position of Homo Habilis and Homo Erectus.

Homo Habilis: The latest find leaves Habilis without any ancestor. Homo Habilis has been a controversial classification since 1960, when Dr. Louis Leakey found the remains of a primitive creature and classed it Homo Habilis (clever man). With its brain of only about 650 c.c., the Homo status has been hotly disputed by many scientists, who have regarded it as Australopithecine.

Some later finds were also classed Homo Habilis, with brains about the same size and smaller.

In 1972, Richard Leakey found Skull 1470, with brain nearly 800 c.c. He declared it human—"a surprisingly advanced early man." (*National Geographic,* June 1973, p. 820). However, opinion was divided on the classification.

He also found leg bones "astonishingly similar to those of modern man" in deposits "older than 2.6 million years." (*National Geographic,* pp. 823 and 828).

Within five years Richard Leakey and his teammate, Dr. Alan Walker, downgraded Skull 1470. It is significant to note that these two experts could not agree on how to classify their own Skull 1470. In a joint paper published in *Scientific American* (August 1978), on page 54 they state: "We ourselves cannot agree on a generic assignment for KNM-ER 1470. One of us (Leakey) prefers to place the specimen in the genus Homo, the other (Walker) in Australopithecus."

It seems that Leakey prevailed, and Skull 1470 entered our lineage as Homo Habilis. All this indicates that Homo Habilis is a classification of merely personal preference. It comprises a motley of creatures, with brains from about 500 c.c. to nearly 800 c.c. It all adds up to a very confused Homo Habilis, devoid of any scientific rigor, and devoid of any demonstrable evolutionary sequence leading to it or from it.

However, let us not lose sight of those human leg bones. Whatever Homo Habilis may be, those leg bones must be disconcerting to the evolutionists' timetable for Man.

Homo Erectus: This seems to be a classification of convenience. The first to be put into it were Java Man and Peking Man, both of which should be regarded as frauds.

We have seen that Richard Leakey, in 1975, classed the Koobi Fora skull as Homo Erectus (KNM-ER 3733), brain capacity about 850 c.c. or 900 c.c. Not long after, he found another skull (KNM-ER 3883) which he put in the Homo Erectus class, though, compared to the first, this skull had more massive brow ridges, facial bones and mastoid processes. Also, I think illustrations indicate a lower, more receding forehead. The 1978 report said that its brain size was not yet determined, but there is no reason to expect it will be much different from 3733 (Koobi Fora). (*Scientific American,* August 1978, p. 55).

Its brain size had not been measured three years after its finding. We wonder why.

Dr. Duane Gish has said: " . . . it is possible that some of the fossils reported as Homo Erectus could actually be Neanderthal Man. Others have told me that some fossils that have been classified as Homo Erectus were classified as such solely because they were believed to be too old for Neanderthal Man." (Letter from Dr. Gish to A.W. Mehlert of Brisbane, dated March 24, 1978).

The Boy: Gish's opinion gains some strength from a recent find by Richard Leakey and Alan Walker in Kenya in 1984, "the most complete skeleton of an early human ancestor ever found." (*New York Times,* October 19, 1984, p. A1.)

This was the almost complete skeleton of a boy about 12 years old and about 5-1/2 feet tall. If he had matured he might have reached 6 feet. It was dated from volcanic ash layers at 1.6 million years old. (Later we will discuss why these dating methods are rejected by creation scientists.)

The *Washington Post* (October 19, 1984, p. A1) proclaimed that the new find reveals that these ancient people had bodies virtually indistinguishable from our own. I wonder, did the *Post* realize that this is not supportive of 1.6 million years of evolution.

The main skeleton looks just like modern man. Dr. Walker said: "It looks so human. I'm not sure whether the average pathologist would notice any differences from a modern human."

As regards the skull, Walker recalled: "When I put the mandible onto the skull, Richard [Leakey] and I both laughed because it looked so much like a Neanderthal." (*Washington Post,* October 19, 1984, p. A11).

"Brain capacity was put at 850 c.c., which is small volume but is within human range."

A: HADAR:	"Lucy" and "First Family of Man" (1974-75) (Combined into "Afar Man"). Johanson and Taieb.
B: LAKE TURKANA:	Skull 1470 (1972). Koobi Fora Skull (1975). The Boy (1984). Boisei (1986). R. Leakey.
C. OLDUVAI GORGE:	Zinjanthropus (1959). Homo Habilis (1961). Louis and Mary Leakey.
D. LAETOLI:	Homo Jawbones, Teeth and Footprints (1974-6). Mary Leakey.
E: TRANSVAAL:	Australopithecus. Dart (1924). Broom (1936).

Apparently, without hesitation, the boy was put in the Homo Erectus class. We suggest that, if the volcanic ash had been "dated" younger, this boy would have been classed Neanderthal, or related to Neanderthal.

From what we have discussed already, there is a growing impression that Homo Erectus is an amorphous classification improvised for convenience. This impression is strengthened by the Australian fossils of Talgai and Kow Swamp.

The Talgai Skull: This was found in Queensland in 1886. It was kept in private homesteads until 1914, when it was acclaimed at a science congress in Sydney. It then lapsed into obscurity until 1948, when N. W. G. MacIntosh, Professor of Anatomy at Sydney University, began years of painstaking work on it.

Science News (April 20, 1968, p. 381), in an editorial, reported: "With its receding forehead, projecting face, prominent eyebrow ridges, and huge canine teeth, [the Talgai skull] has many of the characteristics of so-called Java Man, *Homo erectus* . . . yet it is the remains of a 14-year-old aboriginal boy who lived and died in Queensland 13,000 years ago, and who belonged to the species *Homo sapiens.*"

The report adds: "The significance of the find is that it indicates that *Homo erectus* and *Homo sapiens* might be the same species."

Kow Swamp: A few years after that report on the Talgai skull was published, A. G. Thorne and P. G. Macumber found remains of over 30 individuals in the Kow Swamp area of Victoria. They were dated 10,000 years old.

In 1972, *Nature* (238:316) gave a detailed account of these remains which are human and yet "display archaic cranial features which suggest the survival of *Homo erectus* in Australia until as recently as 10,000 years ago."

In his new book, Dr. Gish discusses the *Nature* report, which states (p. 319) that the skulls indicate "long term preservation of early *sapiens* characteristics" (presumably meaning Neanderthal characteristics), but also that "the frontal bones are particularly archaic, preserving an almost unmodified eastern *erectus* form, specifically that of the Javan pithecanthropines" (i.e., Java Man).

The general skeleton is not described, so Gish assumes it is modern.

Earlier we discussed the finding in Kenya of the skeleton of a boy which Leakey and Walker promptly classified Homo Erectus. Gish suggests that that supposedly erectus skeleton may turn out to be similar to the people of Kow Swamp. (*Evolution: The Challenge of the Fossil Record*, Gish, pp. 202-203).

The Talgai lad and the Kow Swamp people were unquestionably Homo Sapiens. But there it is again: Sapiens Neanderthal and non-Sapiens Erectus in the same individuals, and the experts are scratching their heads.

All things considered, it seems that Homo Erectus (like Homo Habilis) is a creature of the imagination of a small band of fossil seekers.

The Forgotten Men: It is appropriate to recall two discoveries of long ago which should have stopped the quest for ape-men before the quest began. Back in Darwin's day, two skulls of modern men were dug up in rocks which were regarded as millions of years old.

The Castenedolo Skull was found in 1860, in Italy, in Pliocene strata. It was the skull of a modern man. Twenty years later, close to the same spot and in the same Pliocene strata, were found the remains of a woman and two children. The woman's brain capacity was found to be about 1340 c.c.

The Calaveras Skull: In 1866, in California, in early Pliocene gravel was found a human skull of modern type. In the same deposits were stone implements of man.

To any unbiased person, these finds showed that modern type humans were living in the Pliocene period, many millions of years ago (on evolutionist time scale). However, evolutionists are not unbiased. Such bones did not fit into their evolution theory. Consequently they were ignored and suppressed. Evolution promoters acted as if the Castenedolo and Calaveras bones did not exist. However, it is worthy of note that men of the eminence of Sergi and Quatrefages accepted this evidence that modern-type man was living many millions of years ago.

It is high time that the Castenedolo and Calaveras skulls received their due publicity. A good account of them is found in *The Transformist Illusion* by Douglas Dewar (1957), and in Malcolm Bowden's book, *Ape-Men, Fact or Fallacy* (1977).

A recent development, which exemplifies evolutionism's growing turmoil, has split evolutionists into two hostile camps. "The virulence of this controversy and the partisan feelings evoked are remarkable." (*The Neck of the Giraffe*, p. 218, quoting Dr. Colin Patterson).

On one side are the *Gradists,* who are Darwinists. They classify species by presumed evolutionary connections. On the opposite side are the *Cladists,* who disregard Darwinian theory. They classify and relate species by measuring the number of shared features, and without evolutionary presuppositions.

In *New Scientist* (May 3, 1984, p. 25), the head of palaeontology in the British Museum, Peter Andrews, explains what this has done to Homo Erectus. The *Gradists* hold that Homo Erectus was one and the same species in Asia and in Africa, and led to Homo Sapiens. The *Cladists* say that Homo Erectus was restricted to Asia (Java Man-Peking Man) and became extinct there. They say that the African specimens were not really Homo Erectus; but that whatever they were, they evolved into Homo Sapiens.

Peter Andrews' article may be intended to clarify the picture, but it actually highlights the muddle of hominid evolution—a muddle with various contradictory alternatives, and each expert chooses what suits his own ideas—a muddle where it is uncertain whether Homo Sapiens evolved separately in Africa, in Asia and in Europe, or evolved once only in Africa—a muddle where it is uncertain whether Homo Erectus and Homo Sapiens are two separate species or are one and the same species.

If the experts do not know what goes where, one may ask what right has anyone to teach evolution as science?

So ape-men come and ape-men go, and nothing lasts for long.

In *New Scientist* (March 26, 1981, p. 805), science writer John Reader points out that not many (if any) fossil hominids have held the stage for long. He says: "By now laymen could be forgiven for regarding each new arrival as no less ephemeral than the weather forecast."

He also tells us (p. 802) something that is not well known. Fossils are mostly so fragmentary as to admit several interpretations, and "The entire hominid collection known today would barely cover a billiard table."

On those few meaningless fragments of bone a small group of fossil hunters, aided by the mass media, has persuaded the world that brutes turned into man.

In all the confusion we would do well to listen to Lord Zuckerman, who was chief scientific adviser to the British Government. He and his team spent many years investigating the hominid fossils by strict scientific methods. In his book, *Beyond the Ivory Tower* (pp. 75-94), he indicates that he can see no real science in the search for man's fossil ancestry, and no fossil evidence of man's evolution from some lower creature. (Also see *Appendix D*, page 182).

Chapter 5

GEOLOGY AND THE DELUGE

Now we should remember that, up to a hundred years ago . . . belief in the Deluge of Noah was axiomatic, not only in the Church itself (both Catholic and Protestant) but in the scientific world as well. And yet the Bible stood committed to the prophecy that, in what it calls the "last days," a very different philosophy would be found in the ascendant; a philosophy which would lead men to regard belief in the Flood with disfavor . . . declaring that "All things continue as from the beginning of the creation" (2 Peter 3:3-6) . . . And so, after eighteen centuries we find at last the ancient prophecy fulfilled before our eyes . . .

(L. Merston Davies
in a paper to The Victorian Institute, 1930)

This quotation is given by Morris and Whitcomb (in *The Genesis Flood*) with an addendum that Merston Davies, with audience interest aroused, went on to make manifest the disquieting conclusion: St. Peter's prophecy began to be fulfilled in our times with the new doctrine of Hutton and Lyell.

Geologists seem to be the most difficult to convince against evolution; yet geologists can reject evolution and still be first-rate geologists. This chapter deals with the two opposing geologies; the two geologies which contradict each other. They are *Deluge Geology* and modern *Uniformitar-*

83

ian Geology. Which of these geologies really explains the earth and time?

Until recently, Christendom believed that the rocks and valleys had been molded by a flood that enveloped the earth, an earth only thousands of years old. That is *Deluge Geology.*

But lately there have come men with a new geology which mocks the idea of Noah's Deluge and talks of time spans that stun the mind. They say the sculpture of earth was done through billions of years, by the ordinary forces of Nature which we see at work today, slowly and uniformly building and eroding, through immense time, until today. That is *Uniformitarian Geology*, or *evolution geology.*

James Hutton (1726-1797) was the father of Uniformitarianism, which ushered in The New Geology. Hutton did not dispense with a deity. However, he insisted that, from the beginning, the only acting forces have been the forces we see at work today, with no cataclysms, and certainly no Deluge. His hypothesis envisaged "Time," enormous time spans, to mold the geologic formation. In his own words: "We find no vestige of a beginning; no prospect of an end."

In the year of Hutton's death, Charles Lyell was born. Lyell would take Hutton's ideas and popularize them so successfully that uniformitarian geology (wedded to evolutionism) would oust traditional Deluge geology, and become a new world view.

Was it on its superior merits that Lyell's geology gained the ascendancy; or was it because it suited the mood of the times? In this chapter we shall try to answer that question.

Sir Charles Lyell (1797-1875) ushered in the new, multimillion year geology. Lyell was an evolutionist at heart long before Darwin's hypothesis appeared. In fact, not only did Lyell prepare the world's thinking for Darwin, he also prepared Darwin's thinking for evolution.

Lyell kept his belief in God, and in some creation, but not in the Biblical account. The Deluge was unthinkable. Noah had held back the science of geology for too long. For Lyell, time became all-important. Through endless ages of time, the same forces we observe today, the same winds and waters and weather and volcanoes have slowly, uniformly built up the rocks and eroded the valleys. Through an immensity of time, they have shaped and sculpted the face of the earth.

Darwin liked Lyell's geology. It provided the enormous time spans he would need for his biological evolution. Darwin had lost his faith in the Bible and in the miraculous. He had lost his once-strong faith in intelligent design in nature. His belief in natural selection ruled out any intervention of God in evolution. Eventually he arrived at belief in spontaneous generation of living creatures from lifeless matter.

Darwin could not have advanced along this fateful path without the concepts of Lyell. From their partnership of ideas there arose and spread a new gospel. It proclaimed that earth and universe had assembled themselves mechanically from chaos; and, finally, life emerged and evolved upwards to man. The new doctrine flowed away from God. In a letter to T.H. Huxley, Darwin referred to it as the "devil's gospel."

Basic to evolution and to the new geology is the geologic column.

The Geologic Column has become scientifically sacred. Yet it has no physical reality. It does not exist in any part of the world. In any one place, you will find one, or two, or a few of these strata, often with strata missing, and often with the theoretic sequence reversed. The geologic column is not a column you can dig through. It is a mental image only. It is an imaginary column put together by correlating and inserting segments of the fossil record from various parts of the world.

If you think that radiometric dating helped to build the geologic column, forget it. The geologic column was devised long before radiometric dating was heard of. There is no way of telling the age of a rock by examining it. In the early 19th century, a canal engineer, William Smith, began a system in England of classifying rocks by the particular fossils found in them. This became the method for assigning ages to rock strata in England and parts of Europe. Soon it was projected to fit all the rocks of the world.

But does it fit the rocks of the world? Only for those with great faith, as we shall see. And was it presumptuous? As if in answer, a surprising admission was made, in 1956, by Ohio State University's Professor of Geology, Dr. E. M. Spieker, himself a uniformitarian:

> I wonder how many of us realize that the time scale was frozen in essentially its present form by 1840 . . . ? How much world geology was known in 1840? All of Asia, Africa, South America, and most of North America were virtually unknown. How dared the pioneers assume that their scale would fit the rocks in those vast areas, by far most of the world . . . ? The followers of the founding fathers went forth across the earth and in Procrustean fashion made it fit the sections they found, even in places where the actual evidence literally proclaimed denial. So flexible and accommodating are the "facts" of geology.

(Bulletin of the American Association of Petroleum Geologists, Vol. 40, August, 1956). (Quoted in *The Genesis Flood* by Morris & Whitcomb).

The Tyranny of Theory: When evolution theory became dominant, the theoretical age of fossils in a rock became the labeled age of the rock. The geologic column was then based on the great assumption that evolution is true.

All over the world, solid facts contradict the geologic column and the geologic time scale. But, to an evolutionist, the column and the time scale are inviolable. Whatever may be the conflicting evidence, the fossils' theoretical age will dictate the age of a rock, even if this means literally turning mountains upside-down.

Moving Mountains: Around the world we find examples of older rocks (dated by fossils) lying on top of younger rocks (dated by fossils). In the great Matterhorn Mountain, the fossils are in the wrong sequence. By evolution geology, this would mean that the mountain is old but the rocks underneath it are young; that the Matterhorn was formed before the rocks under it. How can they explain that? They explain it this way:

The enormous Matterhorn must have been uplifted somewhere else by an upthrust and then pushed horizontally across country for up to 60 miles. There is another famous peak in the Alps, the Mythen Peak, which is supposed to have been thrust all the way from Africa to Switzerland. Because the fossils are in the wrong theoretical sequence, mountains must be up-ended rather than that the theory be queried.

Rock overthrusts on a small scale are common enough. Under pressures of faulting and folding, small rock formations can slide over other rocks for small distances. These small slides against friction cause crushing and breaking of rock and gouging of rock faces along the contact plane. They leave trails of evidence: layers of broken and powdered rock many feet thick, and gouge marks. A small rockslide is one thing, but moving mountains across great distances is something infinitely different. Common sense ridicules the idea, and science debars it. By engineering mechanics, there is a limit to the size of the moving rock. If the rock is more than modest size, friction resistance becomes so great that the force required to move the rock would exceed the rock's own cohesive strength. The rock would simply break up instead of sliding.

In the case of the Matterhorn and Mythen Peak, no conceivable natural force could move them across country. Even if there were available a sufficient force to move them, those mountains would crack and sunder under that force before they ever moved.

The Lewis "Overthrust": In spite of common sense and logic and engineering principles, evolutionists insist that even bigger mountains be moved rather than that the theoretical age of a fossil be doubted.

In the U.S.A. in Glacier Park, Montana, a small fossil was found (the Coelennea) in a mountain system. To the evolutionists, that dated the upper mountain as Precambrian; but the lower mountain is dated Cretaceous by its fossils. So the upper is dated 500 million years older than the lower. In other words, the fossil story is that the upper mountain was formed before the lower. To solve that dilemma we are given the great Lewis Overthrust. The enormous upper mountain is said to have slid over the lower parts for 35 to 40 miles. As Dr. Clifford Burdick, consulting geologist, puts it:

> In order to get them (the fossils) in the right order, evolutionists propose that a block of the earth's crust 300 miles long and 25 miles wide and 2 miles in thickness, was turned upside-down . . . If there was such movement there are three criteria to look for. One is there should be ground-up rock between the two layers, like the upper and nether millstone . . . Another criterion is breccia, i.e., broken up rock fragments . . . Then there are slickensides (gouge marks) . . . We spent years in efforts to find such evidences at the contact points in Glacier Park, and we have yet to find one single case of contact. We are forced to say that the rocks in Glacier Park are in the same order in which they were laid down . . . not millions of years ago, but thousands of years ago.

And Dr. Burdick added that such large rocks will break up before they start sliding:

> Altyn dolomite does not have the inner strength to withstand pressure, not even local pressure . . . Think of the massive power needed to push a mass of billions of tons of rock over lower rock. The crushing strength of rock had been exceeded many times over and it is incredible to me that evolutionists will hang on to the idea of thrust fault just to save evolution concepts. (From supplement to *Bible Science Newsletter,* July, 1979).

Heart Mountain Thrust (Wyoming): This interesting example is discussed by Morris and Whitcomb in *The Genesis Flood.* It occupies a rough triangle, 30 miles wide by 60 miles long. It consists of about fifty separate blocks which the fossils date as Palaeozoic. Underneath are rock-beds dated Eocene, 250 million years "younger" than the mountain blocks above them. To make the fossil story right, evolution geologists insist that the Heart Mountain blocks were slid into place by gravity from another region. But there is no ground-up broken rock between the two beds. Worse still, there is no conceivable explanation of how 50 separate blocks of rock could have slid there block by block.

Finally, the thrust theory is ruled out by another fact: there are no sourcebeds from which the mountain blocks could have been broken off and transported.

The Empire Mountains (Arizona): These occupy about 30 square miles and are selected as our final example of alleged thrusting.

About 5,700 feet of Palaeozoic mountain (fossil-dated) rests on rocks dated Cretaceous. That would mean the lower rocks are over 100 million years "younger" than the mountain above. The usual overthrust explanation is invoked.

The physical impossibility of thrusting is made clear by Dr. Burdick and Harold Slusher (assistant professor of geophysics, University of Texas, El Paso), who both researched the alleged thrust fault. There was no evidence of ground-up rock or gouge layers. Even more important was the erratic contact plane itself. In places it "resembled the meshing of gears. There could have been no sliding without grinding off the intermeshing projections . . . The top layer fitted the bottom one like a glove, or as melted metal fits a mold."

When the fossils say that mountains around the world are upside-down, the new geology forces the rocks to fit the theory.

I am sure the honest geologist knows in his heart that those mountains did not move.

So, too, the honest geologist must wonder at the innumerable contradictions of the geologic column found in his field work. For instance, he must be perplexed by what is called *Interbedding*, or repetitive strata. Dr. Burdick refers to white limestone followed by a darker band of sandstone or shale, then another band of limestone, until the entire exposure will resemble the American flag. Such exposures occur in Topanga Canyon, California. Along the Alcan Highway from Canada to Alaska, geologists can observe as many as 150 such alternations. All over the world this sort of thing is common, inexplicable under Uniformitarian geology, but no problem to Deluge geology.

The problem deepens when the fossils alternate with the alternating strata. Two notable regions where fossils repeat and alternate are the Highlands of Scotland and the Alps. The interbedding shows normal, conformable deposits of the strata; but the fossils alternate and repeat themselves, thereby breaking the intrinsic rules that fossils should follow.

The Grand Canyon is about the best exposure of the fossil column available. One can look at the earth's crust to a depth of a mile. It reveals astounding strata lines: flat-lying bands of sedimentary rocks, thousands of square miles of horizontal strata. The uniform horizontal lines of this spectacle of strata upon strata proclaim that this vast formation could not have formed slowly, over millions of years, while being shaken by upheavals and subsidences. The facts do not fit the official story.

Could that small river at the bottom have carved this awesome gorge? Surely not! From the river, rising to 1,000 feet, are steep cliffs of Pre-Cambrian rock. Then, at the level of "The Great Unconformity," the Cambrian Muav begins. Here, a sign on the track informs you that "500 million years are missing here." You are being asked to believe that 15,000 feet of late Pre-Cambrian rock formed there, during 500 million years, but somehow it eroded away to a remarkably flat and level surface, ready to receive the deposit of Cambrian Muav which is there now.

The Muav stratum is followed by Mississippean Redwall. This means that (according to the fossils) the Ordovician and the Silurian and most of the Devonian strata are missing. That means a missing 150 million years of geologic column to be explained.

Worse problems follow. The Redwall reverts back to Muav; then the Muav is followed by Redwall; and so they alternate—unpardonable fossil behavior. That alternating series is followed by *another* such series, with Redwall changing to Supai (50 million years missing) then back to Redwall, then more Supai.

The Grand Canyon is geology's showpiece, but it is actually a hostile witness to the geologic column.

(Reference: *Bible Science Newsletter,* November, 1975: Article by Ed Nafziger, who guided tours to the bottom of the Canyon for seven successive summers.)

The Grand Canyon

The Rocks

Modern geology says that the rocks of the earth have been built up gradually, through eons of time, by the same forces which are working today, steadily entombing and fossilizing specimens of creatures of each era.

Deluge geology says that a cataclysmic worldwide flood reworked the surface of the earth, and that there were great volcanic and tectonic upheavals during and after the flood. It says that the fossils represent, for the most part, creatures engulfed by the flood sediments during a short period of a year or so.

A look at the rocks themselves will show that the forces at work today are puny when compared to past forces which built the rocks.

Igneous Rocks, such as granite and basalt, were formed by lava flows. One question for modern geology is: why is there no granite forming today? Another problem is the monstrous magnitude of the volcanic rocks in comparison to the inadequate volcanoes of today.

The world was shocked by the devastation caused by recent eruptions of Mt. St. Helens in America. However, these eruptions fade to nothing when we understand what happened in that region in the past. The Columbia Plateau (in Washington, Oregon, Idaho and California) is built of volcanic material, in some places more than 5,000 feet thick. One layer of this formation is called the Mesa basalt. Geologist Stuart Nevins (in *Speak to the Earth,* pp. 211 *et seq.*) discusses the Mesa layer. It was formed in one single lava flow. It covers about 100,000 square miles and its depth averages about 30 feet. He says that this catastrophic outpouring of the Mesa basalt very possibly took only a few days.

In a footnote (p. 251) Nevins draws attention to some other basalt formations and quotes authorities for his statements. One, in the Parana Basin of Brazil, probably covered at least 375,000 square miles to a depth of up to 2,000 feet. The Deccan basalts of India must have had an original extent of not much less than half a million square miles with an average thickness of about 2,000 feet. These, of course, are the result of continued volcanic activity, not a single eruption, but their scale dwarfs anything that we can conceive by present-day volcanism.

Sedimentary Rocks were formed by water-borne sediments—loose sand and silt and other materials suspended in water, and deposited (or dumped) in layers—which then solidified into rocks.

We have only to look at the Grand Canyon's colossal sedimentary rocks and wonder. What rivers, what floods, even tidal waves, that we comprehend or can imagine; what waters within our experience could transport even a fraction of the sediments needed to build up that massive formation? Yet Grand Canyon sedimentation is small compared to many other sedimentary deposits, such as the Green Shales of Brittany (17,000 feet), the Cuddapah Series, India (20,000 feet), and the Belt Series of North America (40,000 feet). (Ref.

D. Dewar, *The Transformist Illusion,* p. 27).

Morris and Whitcomb (in *The Genesis Flood,* p. 270) make the point that at least three-fourths of the earth's land area have sedimentary rocks in strata varying from a few feet to 40,000 feet or more. Well may we ask, how did such mighty sediments get transported and deposited? Certainly it was not by rivers and floods of the magnitude that we know today. Furthermore, when strata are analyzed, the mystery often deepens inasmuch as normal river action, or local floods, could not effect the peculiar sorting and grading of the deposited materials.

The magnitude of the sedimentary rocks, the peculiarities of their structures, and their worldwide distribution, all bear witness to one-time water action on an inconceivable worldwide scale.

Then, there are the *fossils.* Sedimentary rocks are the only rocks that contain fossils. They are rich with countless fossils all over the world. But, today there is virtually no fossilization going on. There must be a reason; and there is. In order to make a fossil it is essential that the whole animal or plant be buried completely and rapidly.

This rarely happens today. Except in rare, accidental burials, the corpse decays or is eaten by scavengers. Therefore, these great sedimentary rocks, rich in fossils, carry the message that something big happened, something completely different from what is presently happening—a gigantic world flood, which deposited the rocks and which buried within those rocks the creatures we now find as fossils.

Polystrate Fossils: These are inescapable evidence nullifying uniformitarian geology and confirming Deluge geology.

It is not unusual to find fossilized tree trunks protruding upwards through several strata of sedimentary rock, often through 30 feet and more of rock. According to modern geology, such rock formed during millions of years, stratum upon stratum. Can you imagine any tree standing there during millions of years while the sediments were steadily deposited and slowly built up around it?

The polystrate fossils are so important, it is hard to understand why they are so little publicized. There is a good account of them by N. A. Rupke of the State University of Groningen, the Netherlands, in *Why Not Creation* (pp. 152-157).

Although polystrate animals are not as common as polystrate tree trunks, Rupke refers to a report by Dutch authorities. In the United States there is found deposition of sand and clays "in which the entire bodies with the skin impression of huge prehistoric reptiles are met with." Rupke says that, if sedimentation had been uniform, these great bodies would need 5,000 years to be covered. Therefore, the report concludes: "The only possibility is that, immediately after the death, the dead body was covered and, as it were, ensiled by a thick bed of sediment."

Rupke's article gives examples of polystrate tree trunks in Germany and France. It mentions that such tree trunks are also found at different levels. For example, at The Joggins on Nova Scotia there have been reported tree trunks at 20 levels distributed at intervals through about 2,500 feet of sediment.

It quotes a striking example reported from Britain. A leaning fossil tree trunk of great size was exposed in a sandstone quarry near Edinburgh. It measured "no less than 25 meters and, intersecting 10 or 12 different strata, leaned at an angle of about 40 degrees." (See also *The Creation Explanation,* p. 49).

These polystrate fossils are common. They are fatal to uniformitarian geology, but perfectly fit Deluge geology. They offer clear evidence that the sedimentary rocks were not laid down slowly, but were deposited rapidly and by an unimaginable scale of water action.

Coal and Oil: The common idea is that coal and oil

required millions of years to form. But polystrate fossils have been found going through two and more coal seams. Worthy of mention is one found in a coal mine at Newcastle, in England: a fossil tree, 60 feet high, with a diameter of 5 feet at base, standing at an angle, and intersected by 10 separate layers of coal. This is proof that the great coal bed was laid down very rapidly.

Some unexpected things have been found in coal, which prove that coal was formed during man's history. In Germany, America and Italy, in coal supposed to be thirty million years old, have been found such things as two huge human molar teeth; the jawbone of a child; an iron pot; also a neatly coiled 8-carat gold chain.

Carbon-14 dating in the Gulf of Mexico dated oil deposits as some thousands of years; and data produced by the Petroleum Institute in New Zealand showed that petroleum deposits were formed 6,000-7,000 years ago. In the laboratory, scientists have converted garbage into oil within one hour—not in millions of years, but in less than one hour.

Pressure and heat are the factors that produced coal and oil—not million-year time. The real facts testify that coal and oil were formed recently and were formed quickly by the dumping and burial of enormous amounts of organic matter by great flood waters. The real facts testify that the supposed 60 million years of the Carboniferous age are a myth.

Lately, a different theory of the origin of oil and gas has been put forward by geophysicist Thomas Gold, of Cornell University, New York. His theory is explained in his article, "Oil from the Centre of the Earth" (*New Scientist*, 26/6/86, pp. 42-46). He claims that hydrocarbons do not necessarily result from burial of organic matter, but that primordial hydrocarbons, oils and gas, are deep within the earth's crust, up towards earth's surface. There this non-organic oil and gas have been trapped in porous reservoirs (oil fields) capped by impervious layers of rock.

The basis for Gold's theory is that hydrocarbons are plentiful on other planets, such as Jupiter, Uranus and others, where there has never been living matter; and they are also found in asteroids and comets.

The theory is logically presented by Gold, so logically that at least one country, Sweden, is doing test drilling on the strength of it.

If Gold's theory is correct, it does not rule out organic matter as a source of oil and gas. Laboratory production of oil from garbage has been done. So if Gold's theory is correct, it would seem that some oil and gas have come from organic material, and some would be the primordial stuff. The organic could be dated by Carbon-14. The non-organic could not be so dated.

As with many other questions, we await the precise answer which will be forthcoming in due time.

Dry Watercourses and Drowned Mountains

Rivers: Everywhere there is evidence that the rivers of the world were once huge and have steadily shrunk to their present size. For instance, the Thames in England was once 300 feet above sea level, then 150, then 90, then 50 and 20. The waters of the Potomac Basin near Washington, D.C., in the U.S.A., once were 265 feet above their present level, then 215, then 160 and 90. The Grand Canyon must have been scoured by waters 20 miles wide. (Ref. *Bible Science Newsletter Supplement,* May, 1975).

Inland Waters are not what they were. The Caspian Sea is shrinking. There was a time when it was at least 250 feet higher than now, and it was apparently joined to the Aral Sea and the Black Sea in one big inland sea. In the U.S.A., Lake Bonneville and the Great Salt Lake were joined in one great lake that covered much of Utah. Its highest strand line was about 1,000 feet higher than today's level. The scorched Death Valley was once covered by Lake Manley. Lake Texcoco in Mexico was 175 feet higher; Lake Titicaca in South

America was 300 feet higher; the Dead Sea was 1,400 feet higher. Lakes in Africa, in Kenya and Abyssinia were much bigger bodies of water, as was Lake Eyre in Australia.

The Sahara Desert was once fertile land, populated, and watered by rivers and lakes. Other deserts were covered by very large lakes, including deserts of Arabia, central Asia, Australia, North America and Patagonia. (Ref. *The Genesis Flood* pp. 313-314).

Mountains: By sedimentary deposits, fossils and other evidence, we know that the mountains on the earth have been under water. Interestingly, Mount Ararat in Turkey, where the Ark is believed to have rested, is a volcanic mountain. The evidence shows that its lava flowed under water at least up to 14,000 feet up the mountain. Above that level the mountain is concealed under permanent ice. There are also sedimentary rocks at 13,500 feet. (Ref. *The Creation Explanation,* p. 50).

None of the above facts fit the doctrines of uniformitarian geology. All of them do fit the concept of a world-engulfing Deluge, and its aftermath of huge river run-offs, and of trapped inland seas and lakes, which have dwindled ever since, as a sodden earth dried out.

Fossil Graveyards: If you doubt the Flood, then consider the great fossil graveyards which are found around the world. In them are extraordinary numbers of creatures of all kinds, jumbled and mixed together.

There are dinosaur graveyards in many parts of the world containing scores of dinosaur remains, piled up and mixed up and buried together. They did not die there—they were violently washed there.

Some fossil graveyards are high on mountains. In Sicily, 4,000 feet up on Mt. Etna, there are two caves crammed with the bones of thousands of giant hippopotamuses. On the island of Malta, there are lions and tigers and mammoths, birds, beavers, hippopotamuses and foxes, all mixed together. One cave in Malta contains so many animals that

Malta's present size would not keep them in food for a week.

In America, there is a death pit in Los Angeles with tens of thousands of specimens of all kinds of living and extinct animals, including elephants. There is a herd of elephants in one pit, yet, when the Spaniards came to America, there was not an elephant to be seen. Also we remember that the Spaniards found not a single horse in America, but horses have been found buried there by the hundreds of thousands. Encyclopedias tell us that horses mysteriously disappeared in America about 2,000 B.C. There is nothing mysterious about it if you accept the Great Flood.

The Cumberland Bone Cave in Maryland contains dozens of kinds of mammals, as well as some reptiles and birds. They belong to all regions, from Arctic zones to tropical zones, and from dry habitats to moist habitats.

At Gieseltal in Germany, more than 6,000 remains of vertebrate animals, and many insects and mollusks and plants from all climactic zones and all geographic regions are found buried together. In some cases, the soft parts of animals were preserved, as well as hair and feathers, and even some stomach contents.

The evidence proclaims very sudden and complete entombment by sediments of a phenomenal flood. It was certainly not a local flood. Its waters must have swept over all regions of the world.

A dead fish always floats and decays, or sinks and is devoured by scavengers. So, what can explain an astounding Californian graveyard of fossil fish six to eight inches long. Their bodies are crammed about five fish to the square foot over an area of four square miles. That would mean more than a billion fish. Their heads are thrust backwards, showing that they had been trapped suddenly and buried instantly.

In Fifeshire, England, well-preserved fish were found in sandstone. More than 1,000 fish were jammed into one square yard. Again, it must have been instantaneous engulfment.

In South Africa, it has been estimated that eight hundred billion skeletons of vertebrate animals are in the Karroo formation.

Greatest of all the fossil graveyards is that of the Arctic around Siberia. It almost defies description. Buried in the frozen Tundra are innumerable mammoths and all sorts of other species, such as horses, lions, foxes and camels. Estimates have run as high as five million mammoths in those graveyards. A few of them have been so well preserved that their flesh could be eaten (mainly by dogs) after thousands of years.

Fossil graveyards cannot be explained by modern uniformitarian geology. They provide dramatic evidence that an era of the world was ended with enormous violence—and by water.

(Ref. *The Genesis Flood,* pp. 151-161: *The Creation Explanation,* pp. 51-52. Recorded address by the late Dr. Shelley of the Evolution Protest Movement).

Flood Geology

Evolution geology rejects the Deluge and gives us the geologic column arranged to fit the theory of evolution. But, suppose the Deluge did happen (and there is much evidence to support its happening): a cataclysmic flood caused not so much by rain as we know rain today, but the "fountains of the great deep," namely the oceans, enormous upheavals of ocean beds, tectonic activity, and temporary sinking of the continental lands, and inundations by the oceans, with continents submerged by the oceans. There is evidence of this.

Furthermore, many competent people believe there once existed a vapor canopy in the upper atmosphere ("the water above the firmament" of *Genesis* 1:7) which was precipitated by the tectonic activity and poured down to add to the Deluge.

If the Deluge did happen, then the geological column of

FOSSILS RECORD THE SEQUENCE OF FLOOD BURIALS

ERA	PERIOD	EPOCH	SUCCESSION OF LIFE
PALEOZOIC "ANCIENT LIFE"	QUATERNARY 0.1 MILLION YEARS	Recent Pleistocene	
	TERTIARY 62 MILLION YEARS	Pliocene Miocene Oligocene Eocene Paleocene	
MESOZOIC "MIDDLE LIFE"	CRETACEOUS 72 MILLION YEARS		
	JURASSIC 46 MILLION YEARS		
	TRIASSIC 49 MILLION YEARS		
CENOZOIC "RECENT LIFE"	CARBONIFEROUS — PERMAN 50 MILLION YEARS		
	PENNSYLVANIAN 30 MILLION YEARS		
	MISSISSIPPIAN 35 MILLION YEARS		
	DEVONIAN 60 MILLION YEARS		
	SILURIAN 20 MILLION YEARS		
	ORDOVICIAN 75 MILLION YEARS		
	CAMBRIAN 100 MILLION YEARS		
PRECAMBRIAN ERAS			
PROTEROZOIC ERA			
ARCHEOZOIC ERA			
APPROXIMATE AGE OF THE EARTH MORE THAN 4 BILLION 550 MILLION YEARS			

the evolutionists must be re-interpreted. It can no longer stand for a record of the progress of evolution.

Instead of an evolutionary development of different forms of life trying to substitute itself for Creation, the geological column can be seen as a record of the sequence in which creatures were buried by the great flood sediments.

At the bottom of the column are the simplest sponges, jellyfish, sea worms, corals, shellfish and trilobites. Now I suggest that they are at the bottom because that is where they lived. They lived at the bottom of the sea. They would have been the first ones to be buried by the great flood sediments.

The free-swimming fishes were trapped later. Then the amphibians even later, because they lived higher up on the level of the land. So the layers of the fossils would record the sequence of the burial of animals in the flood sediments.

After a certain stage of the flood another factor would come into play: streamlining. The simpler or lower animals are streamlined in water whereas higher, more complex animals, offering more resistance, are not. In moving flood waters, hydraulic principles of streamlining sorted out the simpler, lower animals and caused them to be buried first and lower while complex, higher animals sank more slowly and were buried higher up in the column of sedimentation.

Finally, a third factor came into play that can be called the escape factor. The higher animals are more mobile. Birds flew to higher ground as the flood waters rose and so, too, such animals as horses, apes, and of course, man himself, all being more agile, more fleet-winged and fleet-footed, fled to higher ground as the waters rose and thereby escaped for a little longer. And so they were the last to drown.

The actual fossil column fits this pattern of events very well.

Reflect on what this means. It means that the great sedimentary rocks were deposited quickly by a great flood, and that the fossil column (for the most part) was deposited IN those sediments by the same flood. It means that nearly all the living creatures of the world were engulfed by that flood and buried in the sediments. The creatures became fossils, and the sediments became rocks.

This would mean that millions of fossils in the rocks are (for the most part) creatures which lived on earth together, at the same time, when the Great Flood struck them, whether they were trilobites or dinosaurs or woolly mammoths or man—in fact, the lot.

This means throwing out evolution geology and its billion-year ages, and we are led back to Deluge geology and a young earth.

Chapter 6

RADIOACTIVE DATING

We have been moving the argument toward an earth which is not awesomely ancient, but surprisingly young. But, you may object, what about radioactive dating? Does it not tell us that the oldest rocks are 4.5 billion years old?

Radioactive dating uses minerals like uranium (decaying into lead) and potassium (decaying into argon). Both methods are based on assumptions that are not justified.

The first assumption is that of a *closed system.* The assumption is that the uranium was there, and the potassium was there, in the rocks, for millions of years in a closed system. A closed system is an ideal system which is not possible in the rocks. In reality, the minerals in the rocks would be subject to contamination, to leaching, mixing, evaporation, and other spoiling factors, which make dating by the minerals impossible.

A second assumption is that radioactive *decay rates* never vary. Decay rates will vary if, for instance, cosmic radiation varies. Cosmic radiation almost certainly has varied through variation in the earth's magnetic field, and also through Supanova explosions in nearby stars. Dr. Frederick Juneman commented in 1972 that the effects of such super-explosions must reset our atomic clocks. He added that that would knock our radioactive dating measurements "into a cocked hat."

The third assumption regards *measuring* the "decay-produced" (or "daughter") lead, or argon-40 gas. It is assumed that none of the "measured" daughter lead, or daughter argon-40, was present when the parent uranium or potassium entered the rock. This is not a justified assumption, especially in the light of what violent things can happen to rocks.

Less than one percent of the argon in rocks is the result of decay from potassium. It is difficult to determine which is radiogenic argon and which is not. Furthermore, the dating method rests on another assumption, namely that the *ratio* of argon-40 (radiogenic) to argon-36 (non-radiogenic) has always been 300 to 1. Space probes have shown that the ratio on Mars is very different, with much less argon-36; while the ratio on Venus is very different the other way, with much more argon-36. This is bad news for Potassium-Argon dating, discrediting its very basis.

When potassium decays, it produces both calcium-40 and argon-40. The observed *branching ratio* is 0.12, but in order to make the potassium dating method conform to the already established uranium dating method, the branching ratio that has been used varies from 0.12 to 0.08. In other words, the potassium method has been calibrated to adjust it to the uranium method.

For the sake of brevity, the Rubidium-Strontium dating method has not been discussed. Similar objections apply to it as to potassium-argon.

There is much creationist literature critical of radioactive dating. The above objections to radioactive dating have been selected from writings by Dr. Melvin Cook in *Scientific Studies in Special Creation,* pp. 79-89, and in *Bible-Science Supplement,* October, 1976; also writings by Dr. Harold Slusher in the Supplement, July, 1980; and by Dr. Henry Morris in *Scientific Creationism,* pp. 137-146.

Discrepant Dating Results: These are discussed by John Woodmorappe in *C.R.S. Quarterly* (September, 1979, pp. 102-129). He gives eight pages of examples of radioactive dates which were discrepant with the ages expected, that is, expected according to fossils and stratigraphy. He makes the point that, although many dates do agree with bio-stratigraphy, it is not possible to know the real number of discrepant dates. Why? Because agreeable dates are published but "those in disagreement with other data are seldom published."

He makes the further point that "any discrepant date can be explained away by claiming some event has opened up the system, while at the same time claiming that the alleged event is not recorded in regional geology . . ."; and he quotes some amazing "explanations" given to excuse unacceptable dates.

He adds: "Probably the greatest violations of all are radiometric dates many times greater than the accepted 4.5 billion years of age of the earth." He mentions a Tertiary basalt dated 10 billion years and a Pre-Cambrian rock dated 34 billion years by Rb-Sr. (See p. 105.) These were "explained away." He refers to frequent Potassium-Argon dates 7 to 15 billion years, also "explained away," though they were up to six to seven times the supposed age of the earth.

For further interesting examples of radiometric "ages" which are (a) anomalous or (b) even contradictory within the same rock, see *The Creation Explanation* by Kofahl and Segraves (pp. 195 *et seq.*).

The Real Test: If there were some rocks of known ages, we could test the validity of radioactive dating on them. There are such rocks.

In the sea near Hawaii, rocks were formed by a volcano only 200 years ago. They were dated by Potassium-Argon at up to 22 million years. Other rocks near Hualalei (Hawaii) were formed by volcanic action in 1801. Potassium-Argon dated these young rocks at 160 million to 3 billion years. By the crucial test, Potassium-Argon stands discredited.

What of the Uranium-Lead method? This method involves a chain of intermediate stages as uranium and thorium decay into lead-206 (+ helium), lead-207 (+ helium), lead-208 (+ helium). A sample rock will typically (not always) contain all these isotopes, as well as lead-204, common lead. Supposing the original parent mineral has spent all its life protected in an ideal closed system, decaying ideally along these chains, then the uranium, the thorium and the several lead isotopes should be observed to be in what is called *equilibrium ratios* among themselves. If they are in disequilibrium, there must have been contamination, which means the system was not closed.

Where the real ages were unknown, though presumed to be old, these anomalies apparently were not noticed. But in the case of rocks which were known to be young, it was obvious that the ages calculated had no relationship to the true ages. (Sidney P. Clementson in *Speak to the Earth,* p. 373).

However, sometimes a number of samples will yield *concordant ages,* and that is claimed to confirm the dating and the dating method. Clementson disagrees. He says that, where samples are taken from common sources, the ratios of the isotopes will, of course, give concordant theoretic ages. What matters is whether the ratios are in disequilibrium among themselves.

Clementson then lists research figures from Russian rocks and U.S.A. research data on rocks from Azores, Tristan da Cunha and Vesuvius. All are *recently formed* rocks. All the isotopes were in disequilibrium ratios from which could be calculated "ages" of anywhere from 100 million to 10.5 billion years for these young rocks. Obviously, if the parent element had ever been pure, it had been contaminated.

One basic assumption in radioactive dating is that, at time zero, the parent element entered the rock pure, with no decay elements. Recent research on young rocks has upset this assumption. As the assumption is basic, this means that the dating is not just marginally wrong, but is totally meaningless. Radioactive dating has been wrecked on rocks of known ages.

There may be one lingering question: Is it possible that the radiometric "ages" might nevertheless indicate an ancient earth. Clementson says: "These ages have no relationship to the age of the earth, because, of course, the various ages computed have varied so widely. Consequently ratios of parent and daughter elements are merely ratios . . ." They are ratios, not ages.

It is worth noting that evolutionists cling to discredited radiometric methods, presumably because such methods give some semblance of validity to the enormous ages they need for evolution. They stubbornly ignore the dozens of other phenomena which give evidence for a young earth and universe, which we discuss later.

Carbon-14 Dating: This method is used on organic remains such as bones and trees, coal and oil. It is limited to short terms of up to about 30,000 years. Actually, there is good authority for regarding it as increasingly unreliable beyond about 3,000 years. The Geochron Laboratory in America refuses to use it beyond 3,000 years, claiming it is unreliable beyond that. When you read that C-14 shows that Aboriginals have been in Australia for 30,000 years, and other such statements, question the dating.

Carbon-14 is produced by cosmic rays and nitrogen high up in the atmosphere. The dating method depends on one thing—equilibrium; on production of Carbon-14 up there and its decay down here on earth being in equilibrium.

Walter Libby devised the dating method in 1946, and received a Nobel Prize. His figures showed, not equilibrium, but a 20% imbalance—20% greater production than decay. This seemed impossible, because cosmic rays acting for only about 30,000 years would result in equilibrium; and surely, everyone knew that cosmic rays and earth and everything are billions of years old. Non-equilibrium would mean an earth younger than 30,000 years. Therefore the 20% figure was ignored; it was called experimental error; and equilibrium was *assumed* to exist.

Facts are stubborn things. The non-equilibrium is real. In fact, it is greater than Libby's figure. It is 38% or more. Therefore C-14 dates need to be corrected and creation scientists apply a non-equilibrium correction.

The method also depends on radioactive Carbon-14 and ordinary Carbon-12 being in a constant ratio throughout time. Such a ratio would be altered if an upheaval of nature, such as the Deluge, affected the production of C-14. Many creation scientists believe that, prior to the Deluge, a vapor canopy existed which cut down the amount of cosmic radiation. This would mean that living things absorbed less C-14 pre-flood. Dating their remains would give a small C-14 reading, which would be interpreted as a great age, according to Libby's "equilibrium" model.

It seems certain that earth's magnetic field was progressively stronger as we go back in time. That would reduce the cosmic rays and thus reduce the C-14. The result would be exaggerated C-14 ages in the past.

To a creationist, allowing for non-equilibrium and for the Deluge (and other factors), the great age would have to be corrected to a scale of lesser age.

Working on the non-equilibrium figure, Melvin Cook, Professor of Metallurgy, and Robert Whitelaw, nuclear physicist, have applied corrected figures for Carbon-14 which put the age of the earth's atmosphere at 10,000 years maximum, and quite likely a good deal less.

30,000 Carbon Dates: Robert Whitelaw, in an extraordinary research effort, has checked 30,000 recorded C-14 datings, applied corrections and plotted them. He found two important things:

   (a) There appears to be a period, about 5,000 years ago, when there existed almost no living things, but abundant life preceding that point of time, and then evidence of increasing life following it. He regards this as evidence of extinction of life by the Deluge.

(b) Everything subjected to C-14 testing has given a C-14 reading, except three megapod eggs from a Philippines cave. The testing included many fossil bones, such as Neanderthal Man, Rhodesian Man, and the supposedly extremely ancient Keilor Skull; Mammoth bones (3,370 years; comment "Incredible!"), extinct animals like Sabre Tooth Tiger, coal and oil and fossil wood; also deep ocean cores of what was supposedly the most primitive life.

Virtually every once-living thing was datable by the short-term C-14 method. Many gave astonishingly recent dates, and all gave reasonably recent dates.

For any faithful evolutionist, this is not only embarrassing, it is impossible. If evolution is true, and if creatures have been living and dying for hundreds of millions of years, then in every 30,000 random specimens tested by C-14, only two should have been datable and 29,998 undatable, Whitelaw's research showed the opposite, 29,997 dated and three not datable for some reason.

What is more, with non-equilibrium corrections and Deluge corrections applied, Whitelaw put the oldest dates at no more than 7,000 years, which he suggests would signify creation time. (Ref. "Speak to the Earth" article by R. Whitelaw, pp. 331-364; also *Bible Science Supplement,* July, 1980).

Moon Rocks: When Apollo brought us rocks from the moon, scientists hoped they would establish accurately the age of the moon and the earth and solar system. Those lunar rocks lying undisturbed from time immemorial, with no ravages of wind, rain or flood, should provide superb subjects for dating. But the moon rocks only confounded the confusion.

The moon is supposed to be 4.5 billion years old. When radioactive dating methods were applied to lunar rocks, they gave an incredible variety of ages varying from two million years to 28 billion years.

Zinj and Skull 1470: Dr. Louis Leakey declined to have bones of Zinjanthropus dated by C-14 on the grounds that they would be too old to contain any C-14. Nor was Skull 1470 tested by C-14. In view of Whitelaw's 30,000 dates, it would be interesting to put both fossils to the test.

To date *Zinj,* the Potassium-Argon method was used, not on the bones, but on the rocks above and below the bones. Note well what happened. Two rock samples underneath Zinj were tested. One gave an age a quarter-million years younger than the rocks above Zinj, whereas it should be older. The other sample was worse. It gave an age of half-a-million years younger than the rocks above Zinj, instead of the other way around. To make things worse yet, another evolutionist authority, von Koenigswald, said that the latter test was inaccurate and that the half-million should have read about one million years younger than the rocks above. If you reverse that one-million year discrepancy, I suppose you get two million years' embarrassment in getting the age of one-and-three-quarter million years which Dr. Leakey gave to Zinj. (Ref. *Evolution and Christian Faith,* David-heiser, p. 294.)

Dating Skull 1470: The Potassium-Argon method was used on volcanic tuff adjoining the skull. The first date, by

Zinjanthropus

Skull 1470

a Cambridge University laboratory, was 221 million years. This was rejected as impossible.

Further samples were tested by the same laboratory and gave an age of 2.4 to 2.6 million years. A sample of the tuff was also tested by a foremost dating authority at the University of California, Berkeley, U.S.A., and gave an age of 1.8 million years.

The announced age of Skull 1470 was 2.6 million years. Since then the scientists at Cambridge Laboratory have tested many more samples of the tuff and have gotten ages ranging from 290,000 years to 19.5 million years.

It is easy to understand why the Potassium-Argon clock has been described as a clock without hands—without even a face.

Footprints in Stone: Evolution and modern geology teach that trilobites were plentiful in the earliest seas. They were something like woodlice, some an inch or two long, and some much bigger. The teaching is that they became extinct hundreds of millions of years before man appeared on earth.

In Utah, in Cambrian rocks, there has been found a human footprint, sandal shod, with fossil trilobites embedded in the footprint. In the same Cambrian formation, other sandal prints were found; also two barefoot prints of children. With each there was a small trilobite in the same rock.

Cambrian rocks and trilobites! Both are supposed to have finished hundreds of millions of years before man began.

In their book *The Genesis Flood,* Morris and Whitcomb discuss discoveries of what are apparently human footprints in various places in America in rocks as early as the Carboniferous Period, supposedly over 200 million years old. The prints give every evidence of having been made by human feet when the rocks were soft.

Evolutionists will not accept the evidence at face value. As one explained:

If man, or even his ape ancestor, or even that ape ancestor's early mammalian ancestor, existed as far back as the Carboniferous Period in any shape, then the whole science of geology is so completely wrong that all the geologists will resign their jobs and take up truck driving. (A.C. Ingalls in *Scientific American,* January, 1940).

So, the geologists evade the evidence with explanations that the footprints must have been carved by the Indians, or made by some unknown amphibian.

The same book, and other publications, report the now well-known amazing finds of fossilized *dinosaur footprints and human footprints* in tracks close to one another, even crossing one another and, in one case, a dinosaur print superimposed on a human print. They are located in the bed of the small Paluxy River near Glen Rose in Texas.

A recent development requires consideration. In 1982, stains in the rocks began to appear, stains which seem to add to many of the "human" prints an outline of three large "toes." This suggests a dinosaur foot, yet unlike any known dinosaur. Until the mystery of this unexplained stain is cleared up, judgement should be suspended on whether these Paluxy prints are human or animal.

Nevertheless, let us look at evidence found downstream which indicates that the Paluxy dinosaur tracks were made recently.

New Evidence was reported in *C.R.S. Quarterly* (September, 1979). In August, 1978, clinching evidence was found that dinosaurs are recent. In the Paluxy River bed, about 200 yards downstream from the dinosaur tracks, in the same rock formation as the dinosaur tracks, there was found a charred branch from a one-time tree. The branch is embedded in the "Cretaceous" rock. Quite obviously it fell, while burning, into the limestone mud before the mud hardened into rock.

The dinosaur footprints would also have been made shortly before the mud hardened, otherwise they would not have remained. Therefore the footprints of the dinosaurs (and man) would have been made about the same time as the burning branch fell into the mud.

The charred wood has been dated by Carbon-14 at about 12,800 years old. A Deluge geologist would apply corrections which would reduce that figure to a corrected age of about 6,000 years. This means that the dinosaur tracks were made only 12,800 (or if corrected, 6,000) years ago.

Whatever way the evolutionist looks at it, he is hard put to deny that this evidence shows that dinosaurs lived recently, and contemporaneously with man. However, I do not expect that even this evidence will result in many geologists admitting that their evolution geology is completely wrong and turning to truck driving.

Chapter 7

CONTINENTAL DRIFT
(or PLATE TECTONICS)

When fossils are found in the wrong evolutionary sequence, we have noted that the evolutionist's solution is to move the mountains across country. Moving mountains are trifles compared to moving continents. A new concept has won favor among the majority of geologists in the last twenty years—the concept of continents drifting around the world.

Text-book illustrations generally start with one super-continent (Pangea) containing all the land, and with the remainder of the world a super-ocean (Panthalassa). The new theory says that, 200 million years ago, Pangea began to break up into the continents which exist today. At different times they started to drift towards their present locations. India was an early starter on its way up to join Asia. Africa broke away early but never traveled very far. Australia was the last to start, about 60 million years ago, taking New Zealand along but losing it later. Antarctica, for some reason, has stayed put.

However, Pangea is not the beginning of the new geologic story. The theory confidently assures us that continents have been drifting around the globe for 2,700 million years, moving on their separate plates, bumping and jostling one another. Late in the process, the pre-existing continents, all of them oddly enough, somehow came together and compacted themselves into Pangea (One Land).

The latest chapter began 200 million years ago. Pangea began to break up into our present continents and disperse.

This scenario was scorned by most geologists for decades. About twenty years ago the skepticism began to change to acceptance, and now the theory has become standard geologic creed. From the certainty of fixed continents to an opposite certainty of moving continents, the changeover in geologists' belief has been so rapid as to cause wonderment.

A greater degree of wonderment comes when we note that a number of creation scientists, after giving a compelling scientific case against the new theory, then concede, to some extent, the possibility of some splitting and divergence of continents, not through millions of years, but during and after the upheavals of the Deluge. In the face of this wonderment there is only one place to turn—and that is to the evidence. Does the evidence support or oppose the theory of continental drift or plate tectonics?

For centuries it had been noticed that the west coast of Africa and the east coast of South America suggested a cozy fitting of shorelines. Three centuries ago, a Francois Placet conjectured that there had been a split of continents in the Deluge. Early in our own century, some geologists seriously proposed continental movement but, until the 1950's, this was generally dismissed as fanciful.

The first breakthrough was the discovery of the underwater Mid-Atlantic Ridge a 40,000-mile long mountain chain splitting the Atlantic and coinciding with an earthquake belt. Earthquakes also follow the edges of some continents and chains of islands. This brought the suggestion that volcanic magma may be welling up through the Atlantic Ridge, spreading and creating new ocean floor. This requires a mechanism for disposing of old ocean floor.

The theory then proposed that not only continents, but oceans also are on moving plates; and that, where moving

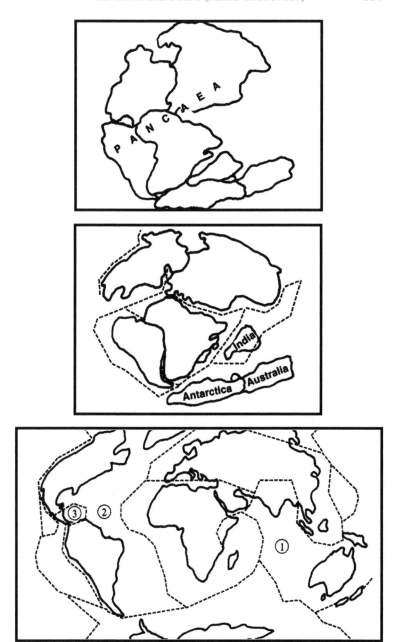

(1) Indo-Australian Plate   (2) American Plate   (3) Caribbean Plate

ocean plates meet moving land plates, the ocean plates are deflected downwards. They are forced under the land plates and down all the way into the earth's mantle. This is called *subduction.*

It is known that the magnetism in some rocks does not line up with the magnetic poles. It was suggested that this could be explained if the continents (not the poles) have moved. Also, from the magnetic patterns of various rocks, geologists claim that the earth's magnetic poles have reversed, flip-flopped, 170 times in the past 76 million years. Next it was found that, on each side of the Mid-Atlantic Ridge, the sea floor exhibits a striped pattern of magnetic reversals which is similar on either side of the ridge.

Potassium-Argon dating was applied to the sea-floor rocks, and indicated that the rocks are youngest near the Ridge, and are progressively older as the distance increases on either side of the Ridge. The magnetic stripes, now dated, made geologists interested. They concluded that volcanic magma was issuing from the Ridge, spreading, and form-ing new sea-floor in each direction. They decided that, as the new sea-floor hardened, it preserved the direction of the earth's magnetic field at the time, and then moved onwards. This convinced them that the *Atlantic Ocean* is opening and widening on either side of the Ridge, at a rate of an inch or two per year. Subsequently, it was estimated that the *Pacific* floor is moving six to eight inches per year.

More good news came with the discovery that, on each side of the Mid-Atlantic Ridge, sea-floor fossils (on the evo-lution time-scale) indicated a sea-floor which is older the further it is from the Ridge.

Then a fossil reptile (Lystrosaurus) was found in Antarc-tica similar to a reptile that had lived in Africa, India and China around 200 million years ago (evolutionist time), a reptile which could not have crossed oceans. It was decided that this meant that Antarctica and Africa were once joined.

More evidence turned up in findings that radioactive datings, and formations of rock strata and minerals along the Gulf of Guinea (West Africa) showed a surprising correspondence with strata and minerals and ages around the eastern corner of Brazil (South America). Also, sedimentary rock formations in West Europe are similar to formations in some places of eastern North America, without similar formations occurring under the intervening Atlantic Ocean.

In the context of evolution, these various discoveries had established a *prima facie* case warranting a close look at the theory. It seems that evolutionists have gone much further than that and, in the name of geology, are speculating freely. The speculation takes on a dogmatism; for example, that pre-Pangea continents had drifted around for billions of years. Also, that the Sahara once lay at the South Pole with Africa 6,000 miles apart from North America; but that, 250 million years later, Africa and North America had come together in a snug fit as part of Pangea.

Is this exact science, or is it similar to the so-called science that tells us that hydrogen gas evolved into man? Before examining the main points of evidence, I make a comment (for what it is worth) that the once-small Australian Plate seems to have changed into the enormous Indo-Australian Plate. Also, the modest-sized Plate that first carried South America has now become the colossal American Plate; with a small Caribbean Plate appearing from nowhere. The suspicion arises that the theorists are tailoring Plates to fit the theory.

Furthermore, the theory proposes that the Red Sea and its connected Great African Rift are probably the beginnings of a new ocean which is just starting to open up. I suggest that geology could just as well explain it all as a massive *subsidence* of rock strata of fixed continents. Likewise, the San Andreas Fault in California, which is claimed to be the edge of the Pacific Plate grinding along the edge of the American Plate, could be as validly explained as the result of rocks faulting, thrusting and settling.

The Evidence: Serious objections to Continental Drift are:
- It requires time-spans of millions and billions of years. I submit that evidence which is cited in Chapter 10 in this book limits the age of earth and rules out any such time spans;
- It depends on dating by fossils. Again, I submit that this book has given enough scientific fact to invalidate fossil dating;
- It depends on Potassium-Argon datings of the sea bed. We have already discussed the fatal weaknesses of long-term radioactive dating, with emphasis on Potassium-Argon. Evolutionists persist in trusting radioactive dating, but surely they should not trust it for dating *sea-floors.*

On the unreliability of undersea dating, geologist Stuart Nevins, in an article for the *Impact Series of I.C.R.,* says that Potassium-Argon dating, when correctly interpreted, shows no evidence of increasing age with distance from the Ridge. His reference is P. S. Wesson (J. Geol., volume 80, 1972).

Dr. H. Morris (in *Scientific Creationism,* p. 147) quotes authorities Noble and Naughton regarding impossible Potassium-Argon ages of Hawaiian submarine rocks of recent formation:

> Ages calculated from these measurements increase with sample depth up to 22 million years for lavas deduced to be recent. Caution is urged in applying dates from deep-ocean basalts in studies on ocean-floor spreading. (*Science,* October 11, 1968).

Morris then reminds us that, although the Potassium-Argon method has been warned against for underwater dating, the modern concept of continental drift is largely based on such dating of the Atlantic floor.
- It relies on magnetic "reversals" and their "zebra stripe" pattern on the supposedly spreading sea-floors:

Stuart Nevins says,

> There are some major problems with this classic
> and "most persuasive" evidence of sea-floor spread-
> ing. First, the magnetic bands may not form by
> reversals of the earth's magnetic field. Asymmetry
> of magnetic stripes, not symmetry, is the normal
> occurrence. (His reference: Meyerhoff and Mey-
> erhoff, *American Association Pet. Geology Bulletin,*
> Volume 56, 1972.)

The question arises: *Can* the earth's magnetic field
reverse? This depends on what causes the field. It must be
an electric current within the earth. Evolutionists say that
the current is produced by a dynamo effect caused by molten
eddies within the core of the earth. This is theoretically conceivable but, I understand, it involves such a complexity of
requirements as to make it extremely unlikely.

In an article, *Pangea Shattered (C.R.S. Quarterly,* June,
1979), Mark W. Tippets says:

> What is not known, however, is how the earth's
> field is generated and if (in fact) the field can reverse
> itself. Mere belief that the earth's magnetic field is
> dynamo-generated, and that it reverses, is not evi-
> dence. It is known that the earth's field has retained
> the same orientation for at least 10,000 years.

The only alternative to the dynamo theory seems to be
that there is an existing electric current, not being generated but decaying. This would produce a decaying magnetic
field. The decaying magnetic field has been observed and
measured during a century and a half. Professor Thomas G.
Barnes has made it a special field of study and advocates
the decaying electric current concept. This would *not* permit magnetic reversals.

What, then, could produce the observed magnetic reversals in the rocks? It seems that some minerals and rocks are

capable of undergoing magnetic self-reversal, but laboratory tests have shown that only a very small percentage have this property. There are authorities who state that several other mechanisms can cause rocks to acquire a magnetic field out of alignment with the earth's field, such things as (a) differences in magnetic susceptibility, (b) tectonic stresses, (c) shock, (d) fractures and (e) chemical causes.

Tippet's article says: "Magnetic reversals observed on the floor of the ocean are real, but their cause is at the present time unknown." He quotes authorities to the effect that on the bottom of the Pacific and North Atlantic the magnetic patterns are sometimes garbled; and that the Great Magnetic Bight in the North Pacific seems to require impossible forms of sea-floor spreading and plate movement; and that there have been found no magnetic lineations on the older part of the floor of the Indian Ocean.

Apparently, much more than is warranted has been made of magnetic patterns and datings, in an enthusiasm to justify the theory of continental drift.

Now we look at evidence which should completely discredit the theory.

Driving Force: The greatest weakness of the theory is that there is no force known, or even conceivable, that could make the continents move. Search the literature and you will find that the geologists, who propose this radical theory, unabashedly admit that the greatest mystery of the theory is *what makes it move.*

Let us soberly contemplate the enormity of hypothetic continental plates, 6,000 miles long, thousands of miles wide, and hundreds of miles deep.

What force could move such enormous masses? What force could not only move them, but also have such additional reserve power that India could be pushed into Asia so irresistibly as to force the mighty Himalaya Mountains to rise up from ground level to their towering heights?

Again, if ocean plates subduct, what imaginable force could drive one huge plate under another and then ram it down 400 and more miles into the earth? The theorists are so puzzled themselves about the motive power that it is fair to predict that they will never find an acceptable answer; and that suggests there is *no* such force and the theory is wrong.

Subduction: Along the Peru-Chile Trench and the Eastern Aleutian Trench it is theorized that one plate is being driven under another plate and bent down into the depths of the earth. If this is true, there should be found along these Trenches sediments that are compressed, and deformed, and thrust-faulted. The contrary has been found. Along the Trenches are only soft sediments without compressed structure. Where we should find piled-up sediments which subduction is expected to produce, we find, instead, flat-lying sediments.

Also, along the island arc and along the coastal mountains on the landward side of the trench (mountains supposedly pushed up by pressure of the subducting plate), these islands and mountains are found to give gravity readings showing crustal material of low density and not compressed.

Stuart Nevins (in another *Impact* article, 1973) says that approximately 27.5 billion tons of sediment are deposited on the ocean floors each year; but those who promote the subduction theory can account for the destruction by subduction of only one-tenth of these sediments. Ocean sediments are forming 10 times faster than they can be gotten rid of by subduction. This is not only strong evidence against plate tectonics, it is also good evidence for a young earth.

The "Jigsaw" Fit: The fitting of one continent into the edges of another can be done in a number of different ways. There are various "fits." They cannot all be right. The best "fit" (the "Bullard fit") has areas of overlap of continents;

and it has one large area, Central America, which must somehow be omitted from consideration.

Also, there are "fits" which are geometrically feasible but
which are preposterous to the theory; for example, rotating
East Australia to fit into eastern North America. (Ref. Stuart Nevins).

An Alternative Explanation

In *Readings in Earth Sciences,* 1972, an article on continental drift by John C. Maxwell (taken from *American Scientist,* Volume 56, 1968) gives a number of geologic
arguments against continental drift, including lack of motive
power and difficulty of disposing of old crust. He says that
the idea of under-thrusting of oceanic crust beneath continental margins, if it is envisioned that these zones are the
expression of thousands of miles of under-thrusting, is highly
unlikely on at least two counts:

Firstly, under-thrusting of this magnitude should initiate
compressive buckling of the crust; instead, the associated
oceanic downwarps seem to reflect passive sinking and
stretching.

Secondly, there is the mechanical difficulty, if not impossibility, of forcing thousands of kilometers of light crustal
rocks downwards into heavier mantle rocks.

Dr. Maxwell also says:

> The effects of ocean spreading seem obvious and
> uncomplicated so long as only a single ridge is con-
> sidered—for example, the Mid-Atlantic Ridge.
> Spreading related to the ridge presumably has car-
> ried the Americas westward and Africa and Europe
> eastward by similar distances. If one considers the
> total pattern of the oceanic ridges, however, they are
> seen to be distributed quite symmetrically—around
> Antarctica, between the Americas and Europe-Africa,

and between Africa-Asia-Australia. This pattern cannot be produced by simple ocean spreading, unless the spreading is a manifestation of an expanding earth, as postulated by Carey (1958), Heezen (1960), Creer (1965) and others. It is more likely that these oceanic ridges are adjusted to the present continental configuration and are a consequence, rather than a cause, of the distribution of oceans and continents.

Dr. Maxwell looks at two sides of the coin:
  (a) whether the forces issuing out of the ocean ridges produce moving plates which eventually subduct under land plates, and the pressure forces up mountains, or
  (b) whether certain similarities between *ocean ridges* and *young mountain systems* (from Gibraltar to Malaya, and then encircling the Pacific), plus the fact that ocean ridges and young mountain systems "seem to merge along trend, suggests a *fundamental similarity.*"

As I understand it, Dr. Maxwell presents arguments that the ridges in the oceans, and the chains of young mountains on land, form links around the globe; and they may be caused by the same forces deep under the earth, namely, rising hot mantle material.

Instead of the ocean ridges producing the cause of these mountains, both ridges and mountains are the *effects of the one cause.* Instead of the continental-drift concept of planet earth being divided into 10 or so moving plates, whose boundaries are marked by belts of earthquakes and volcanoes, Dr. Maxwell's argument leads to something quite the reverse. In fact, it is so much the reverse that, perhaps, it wraps up the argument against continental drift:

We thus have a globe in which both oceanic ridges and young mountains are the expression of diapiracally rising hot mantle material, forming the margins of ten "cells" of roughly equal size distributed over the surface of the earth.

There still remains a minority of outstanding earth scientists (Jeffries, Meyerhoff, the Russian geophysicists, *et al.*) who oppose the idea of continental drift as geophysically impossible, and there are some signs that indicate the pendulum may be starting to swing back again. (*Scientific Creationism*, p. 128).

Chapter 8

CULTURAL EVOLUTION OR DEVOLUTION?

The present size of the world's population should suggest that mankind cannot have inhabited the earth for unlimited time. In this light, several authorities have examined population growth, some with the aid of computers programmed to allow for difficult factors.

Results vary, but the overall picture is clear. Allowing a rate of increase much smaller than today's (and starting from one family, say Noah's), our present world population figure would be reached in roughly 5,000 years or less.

Such figures are disastrous for the evolutionist claim that man has been on earth for at least a million years, which would mean at least 25,000 generations. Dr. Morris calculates that an average family size of 2.5 children during 25,000 generations would produce a present population of 10^{2100}. This fantastic population figure can be appreciated when we compare it with the number of electrons in the entire known universe, estimated at only 10^{130}.

Population statistics make the time span for man's evolution look absurd, but support the Biblical time span.

The popular modern idea is that the human race spent a million years, or more, slowly improving itself from primitive level to high culture. This idea runs into a bad problem with stone-age men. The problem is this: Why is it that intelligent stone-age men scarcely improved themselves at all?

Stone-age men had all the intelligence that we have. Except where a high culture is on the decline, man will improve his culture rapidly. Stone-age man (according to the evolution story) was man on the way up. How then could these highly intelligent men have gone on and on, scarcely improving their cultural level at all, not just for thousands of years, but for hundreds of thousands of years according to evolutionary time?

Evolutionists say that the human race began with many brutish first men and women coming from animal parents, and that from these brutish beginnings mankind slowly, over enormous time spans, climbed the cultural ladder. On the other hand, the Bible says that man began when Adam and Eve were created in a high state, and that mankind was civilized and cultured from the start.

History and archaeology support the Bible account that mankind's story is a short one. And only a few families lived in caves, as some do today. Most people then, as now, lived in cities and towns. These ancient people were highly skilled and cultured, perhaps more so than our own present culture.

The archaeological record shows this: the Middle East was the cradle of civilization and there never was a savage in that area. In this cradle, not many thousands of years ago, starting from a primitive settlement near Mount Ararat, there quickly flowered on the plains of Mesopotamia the great prehistoric civilization of Sumeria—a high culture from nowhere.

According to the testimony of the spade, culture starts, ready-made, in the very center of the world—the Middle East. All radiations of culture are from this central cradle.

The subject of prehistory is deep water, confused water, and authorities conflict. An amateur like myself must never be dogmatic. Nevertheless, following the views of a number of scientists and scholars, this is the interpretation that I have arrived at:

Man was civilized from the start. The story of man is divided into two parts by a cataclysmic flood. The cradle of all culture was in the center of the world—the Middle

East—both before the flood and after the flood. Then, as men migrated away from the center of culture and the mainstream of tradition, so did their culture deteriorate.

According to this interpretation, for which there is very impressive evidence, the post-flood era begins at Mount Ararat where the Ark is believed to have grounded. When Noah stepped out of the Ark, there were only eight people alive in the world, and they were eight cultured people. They had the culture and the technology of a civilized world that had just been wiped out.

Probably Noah's descendants lived in the highlands for a time. Then it seems that some of them moved down the east side of the Zagros mountains to the first settlements of Iran, and then to Susa, then on to the plains of Mesopotamia, where the great civilization of Sumeria quickly arose with a culture ready-made, in a land which had previously contained civilized people before the flood.

It is impossible to fix dates before written history, but Sumerian civilization flourished perhaps 3,000 to 4,000 years before Christ, and it did *not* evolve *up* from savagery. There never was a savage in this land. When we look back 50 centuries to Sumeria, the curtain rises on a people fully civilized, living in prosperous cities and using metals. The high point of Sumerian culture was its early period with artistry that amazes us with its beauty.

Post-Flood Ur was a great Sumerian city. It was the city where Abraham lived. From the archaeological work of Sir Leonard Woolley, we know a great deal about Ur, both post-flood and pre-flood. This high civilization appeared from nowhere. The artifacts and the dress of these people, recovered by archaeology, manifest a degree of culture which matches any existing today.

The Sumerian culture was carried to Egypt. There arose the mighty Egyptian empire. Then migrants from this central area moved to India and the great Indus civilization was founded. China, before 2,000 B.C., was being peopled by

civilized men from the central cradle. To Britain, around
2,000 B.C. came megalith men, the great stone builders, by
way of Brittany in the northwest of France. These men
brought with them an impressive technology and organization which we are only now beginning to realize and fully
appreciate. A thousand years before the glory of Greece,
megalith men were building monuments to their genius like
Avebury and Stonehenge. Megalith man was an expert engineer, geometrician, meteorologist, astronomer and boat-
builder. Some of his precise workmanship is surpassed today
only in the most specialized types of surveying. From that
culture came a decline to the Britons that the Romans found.

Crete received its culture from Egypt. From Crete it passed
to Greece and from Greece to Rome. Through all these links,
the culture which appeared in Sumer (near Mount Ararat)
has been handed down to us.

Here is an observation: A culture does not automatically
evolve up or devolve down. It flowers from a fusion of three
enlightenments: (1) spiritual, (2) intellectual, and (3) tech-
nological. If one of these three declines, the whole civilization begins to decay and decline.

It is popular to talk about culture evolving *up from* savagery. But the evidence shows the reverse. It shows culture
degenerating *down* to savagery; this means that primitive tribes
in the modern world are not man on the way up—they are
sad evidence of man's tendency to go down culturally.

Let this be noted well. There is not one example in history where a savage people, left to itself, has civilized itself
or even improved itself.

On the other hand, whenever a high civilization has fallen,
its descendants, if left to themselves, have never risen up
again to their former high level.

According to the evidence, the story of man starts with
a high culture around the Middle East; then zones of
descending culture radiate further out. As men migrated outwards from the center, or were pushed out, and as they

became more cut off from the mainstream of tradition, so also was their culture reduced. So we find descending grades of culture right out to Tasmania. Examples of degeneration of culture can be quoted from many places, the result of remoteness from the mainstream.

The most remote, the most isolated, the Tasmanian aborigines have been said to rank near the bottom of the cultural scale.

But this is not yet the bottom. There is another stark proof that man, when left to himself, will not rise but will sink—even to animal levels.

Feral Children: Over the past couple of centuries or so, certain wild children have been captured. These are children who had been abandoned in infancy but somehow survived in forests, cut off from all human contact. All of them are so similar, so savage, that Linnaeus could not conceive that they had been born human. He regarded them as a separate species. He classified them as "homo ferus," wild man. They were all strange, seemingly half-witted creatures with no interest in what went on around them. They had the organs of speech, but they seemed unable to be taught to speak.

Some of the many reported cases are well authenticated; for example, Peter, the Wild Boy (near Hanover, 1723); Victor, "The Savage of Aveyron" (France, 1799); also two little girls, aged about 8 and 1-1/2 years, Amala and Kamala, who had been nurtured by a she-wolf (India, 1920).

Victor was about 12 years old. Patient effort was spent, but he could be taught only a few single words, and they seemed to mean nothing to him.

Amala and Kamala had each other, but there was no childish chatter between them. They did not speak. Amala, the younger, died. Kamala, in a civilized environment, was almost incapable of conversing. After six years she had learned about 40 words, could make sentences of two or three words, but never spoke unless spoken to. (See Appendix C.)

Feral children make manifest human nature deprived of cultural teaching. Humanity has not that something within its genes whereby, unaided, it can raise itself up culturally. It needs a culture gifted to it, infused into it.

Wild children highlight the *mystery of human speech*—the great chasm between man and beast. Man is unique. By language, sounds and symbols, men exchange information and ideas, material and abstract.

The theory that men evolved from animals led to the theory that animal grunts and cries evolved into human speech. Such a theory assumes a natural progression from animal sounds to human words. The feral children have proved this wrong. They prove that man, in the company of animals, remains dumb, with no instinct or inclination to utter any word. If speech and language really did evolve, we should find, among primitive tribes, languages in various stages of imperfection. Instead, and to the bafflement of the theorist, it is found that every race, every tribe—cultured or savage—has its perfect language.

Savage tribes may lack all other marks of culture, but they never lack a perfect language. In her book, *Philosophy in a New Key,* Susanne Langer remarks in wonderment that the most primitive of people, lacking textiles, mindless of filth, who even roast their enemies for dinner, "yet will converse over their bestial feasts in a tongue as grammatical as Greek and as fluent as French."

Linguistic research, and experiments with animals have shown the unbridgeable gulf between animal noises and human speech. Dr. Henry Morris has quoted Dr. Noam Chomsky, one of the world's foremost linguists, on the question of whether the lower form of animal noise could evolve into the higher form of human language:

> There is no more of a basis for assuming an evolutionary development of "higher" from "lower" stages, in this case, than there is for assuming an evolutionary development from breathing to walking.

In other words, animal sounds and human speech should not even be considered as comparable functions.

Language is an insurmountable problem for the evolutionist. It presents no problem for the creationist—the Tower of Babel explains it all.

Underlying language is the *deep structure* of thought and reasoning which is universal among mankind, but is also unique to man. Without humanity's criterion, language, the feral children's behavior was not human. Their resistance to learning indicates that the gift of language, once lost, is more or less irrecoverable. Yet, the organs of speech were there and so was the brain formation. Perhaps it is the *deep structure* of thought and reasoning which atrophies if deprived of the catalyst of culture in early years.

Whatever happened inside these children, it tells us that man does not evolve culturally; that man was designed to have a culture as a birthright; and that man must have a teacher. But there was no man to teach the first man—no one, except his Maker.

"Fluent as French"

Chapter 9

LIFE

Only life begets life.

The work of Redi, Pasteur and others settled a long dispute. They showed that life is not spontaneously generating on earth. That has posed a problem for evolutionists: how to get life started on earth without a supernatural agent.

While the problem still remains, its difficulty has been bypassed. The public has simply been told that there is no problem. The great channels of information have tirelessly carried the message that life generated itself by natural chemistry. Thousands of good scientists may disagree; but, generally, their dissenting views are withheld from the public. In this way, the modern media has made people believe as fact that which is impossible.

I dare say that the man of the world visualizes first-life as a blob of jelly, different from other blobs because it was lucky enough to acquire life. This is a misconception. Strictly speaking, there is NO such thing as living matter. This will become clearer as we proceed.

The electron microscope has given some insight into the staggering complexity of a living cell, and some understanding of the interdependence of its various parts, each part interacting precisely with multitudes of other parts in a wondrous organization.

In the heart of the cell is the *nucleus.* This is the control center, masterminding the operations of the cell through prodigiously complex molecules of *nucleic acid* (DNA), and the genes which make up those molecules.

Genes are units of heredity. Each carries the genetic code for some characteristic of the body. The code is spelled out by hundreds of smaller units called *nucleotides,* arranged in highly specific sequence within the gene.

Chromosomes are strings of genes, and the genes are strung in precise and specific sequence. The chromosomes form paired arms, twin arms, which coil around each other in a double spiral. In the human cell there are 46 chromosomes arranged in 23 pairs along the twin arms. In the nucleus of any cell, the chromosomes contain the coded blueprint for structuring the body.

The Double Spiral

Membrane encloses the nucleus, and through it chemicals can come and go. On the other side of the membrane is a fluid called *cytoplasm* in which countless tiny bodies carry on the never-ending business of the cell.

An outer membrane encloses the cell, and it has the ability to decide which materials may enter the cell and which may leave it—a mysterious process which baffles science.

Building a protein: Inside the cell there is continuous activity building new proteins. The blueprint for each type of protein is coded in a gene in the chromosomes. Protein building begins when an extremely complex enzyme (RNA *Polymerase*) selects the appropriate gene. The enzyme receives a "Start!" signal. It sets to work scanning the gene, and it builds an RNA-molecule in the image of the blueprint. When the job is done, the enzyme receives a "Stop!" signal. The new RNA-molecule takes on the role of *Messenger-RNA,* and it goes forth bearing its image, or copy, of the code.

It goes through the membrane into the cytoplasm, where

it is captured by one of thousands of *Ribosomes,* which are complex beyond our understanding, and which are the manufacturers of proteins. The ribosome's task is to build a particular protein by linking various amino acids in the specific sequence of the blueprint. For this task it needs many slaves to catch the different amino acids. The slaves are bodies called *Transfer-RNA,* and each is designed to catch a particular amino acid. They need subservient slaves, special *enzymes* which are suitably shaped for driving one type of amino acid on to the equivalent transfer-RNA body.

The ribosome receives a "Go!" signal and begins to scan the copy of the code carried by the captured messenger-RNA. The slaves muster and catch the various amino acids and put them at the disposal of the ribosome. Working swiftly and silently, the ribosome links the amino acids, about 30 per second, in the specified coded sequence, and *builds the new protein.* When the protein is complete the ribosome gets a "Stop!" signal. The new protein is carried off to do its work in the body.

Such an over-simplified description does not do justice to the intricate operations involving many other steps and numerous specialist enzymes. There are hundreds, and even thousands, of ribosomes at work making all kinds of proteins to specifications. All around are other busy bodies doing their special jobs precisely in tempo with one another; the *Nucleolus* which manufactures ribosomes; the *Mitrochondria* supplying energy; the *Golgi* bodies packaging and disposing; the *Endoplasmic Reticulum,* a double-layered membrane dividing the cell into compartments; *Regulatory Genes* which cause *Operator Genes* to be switched on and off; the *Anti-bodies* of various design which grab and destroy invading poisons and germs; and on and on.

DNA is needed to make enzymes; but enzymes are needed to make DNA. Which came first? This is a tough problem for evolutionists. Operator genes would wreck a cell if the Regulator genes were not controlling them; but Regulatory genes would be purposeless without Operator genes. So, which came first?

The living cell is the most complex structure that exists and its every part depends somehow on its other parts. When it is realized that a cell could not evolve part by part, but must exist in toto or not at all, then it becomes clear that the propaganda about spontaneous generation of first life is unscientific wishful thinking.

This brings me back to the statement that there is no such thing as living matter. In the cell, all the parts and pieces are separate units, working precisely together, but not themselves alive. Life is not IN any of those parts and pieces, just as, in your automobile, there is no automobility in any piece or part, in a spark plug or a carburetor needle. Automobility is a super-quality which coheres to the total motor car. Similarly, life is a super-quality that coheres to the total cell.

Suppose from a living cell we extract any structure. Suppose we extract a ribosome, a membrane, or any protein, even some DNA, we will find that such structure, when separated from the total, organized cell, is nothing more than an organic molecule—a lifeless molecule. Whatever life is, life coheres to the organized cell, the complete cell.

In the fantastic processes within the cell, amino acids are nothing more than humble building-blocks used by the master builder. This has to be kept in the front of our minds because, when Dr. Stanley Miller's famous experiment in 1953 showed that amino acids can be made by natural chemistry, sweeping claims were made that the mystery of life was nearly solved.

The background to the experiment was the realization that free oxygen in earth's atmosphere would prevent the spontaneous formation of amino acids and other organic compounds. With typical resourcefulness evolutionists hypothesized a primordial atmosphere without free oxygen, but with all the ingredients most suited to produce what they wanted. Miller put the hypothesized gases into a container, a mixture of methane, ammonia, hydrogen and water vapor. He subjected the mixture to prolonged electric sparking which represented flashes of lightning. The resulting products were separated and collected in a *trap,* and they were found to include some types of amino acids. Here were amino acids formed by natural processes. Evolutionists proclaimed that this demonstrated that natural processes could produce the pre-life "soup" out of which life could emerge.

The first point to remember is that amino acids are very humble factors in the structure of life. On other points the evolutionists' claim is easily refuted:

- The earth's atmosphere almost certainly never at any time resembled Miller's mixture.
- Even if we presume an atmosphere like Miller's, if any amino acid had formed on earth it would have been immediately destroyed by the same source of energy that formed it—because there would be no oxygen to shield it. The trap which was provided in Miller's apparatus protected the amino acids from destruction by the same electric sparks that formed them.
- Amino acids have an intriguing characteristic; they can be either left-handed or right-handed. The only difference is that a solution of left-handers will cause polarized light to rotate in one direction, while the right-handers cause it to rotate in the opposite direction. The second law of thermodynamics requires that randomly formed amino acids consist of equal ratios of left-handed and right-handed molecules. Miller's amino acids were a law-abiding lot; they were a correct mixture of left-handed and right-handed molecules. The catch is that LIFE goes outside the law.

In living cells the amino acids are all left-handed—left-handed only: never a right-handed molecule. This is one of the fascinating mysteries of life, revealing, not blind chance, but an imaginative intelligence at work.

So, the living cell breaks through the laws governing inanimate nature. Life, then, must be beyond these laws. This point is doubly proved in *death*. At death, the law of non-living matter takes over. The amino acid molecules start to rearrange themselves slowly until, eventually, they are in equilibrium, with an equal ratio of left-handed and right-handed molecules in the cell from which life has departed.

Among other reasons for rejecting the claim that experiments like Miller's have helped solve the riddle of life, there is a neat and simple one. The pre-life "soup" would have to contain amino acids and sugars; but they would not follow the scenario. Amino acids react readily with sugars to form compounds useless for life. The predominant amino acids would render the sugars useless. Yet, both are needed for life: amino acids to form proteins, and sugars to help form DNA and RNA.

Furthermore, producing a few amino acids is child's play compared to the production of vast quantities of those that are biologically active, in their correct relative volumes; and then, there is the superhuman task of linking them together in the strictly precise sequence needed to form a protein.

Much prominence has been given to claims by Sidney Fox that he has formed chains of amino acids into something like proteins. He uses absolutely *dry* amino acids (pure, with no other substance present) mixed in artificial ratios. He heats his pure and dry amino acids at 175°C for six hours. He then stirs with hot water and filters out impurities. When the solution cools, chains of amino acids precipitate out. Fox calls them proteinoids.

Actually, they are nothing like proteins. They are mixed left-handed and right-handed amino acids. They are strung together without the specific order which a protein would require. They are of shorter length than protein chains, and they are floppy chains which could never contribute to the production of life.

Regarding the requirement of high temperature (175°C)

for six hours, Fox suggested that the edge of a volcano would suit, provided that a shower of rain came at the end of the six hours and washed them away. This seems to be rigging the requirements. Temperatures much higher than 175°C are attained at the vents of volcanoes. Also, volcanic gases are water-laden, thereby defeating the requirement of dry amino acids.

Fox's process requires amino acids in artificial ratios which would not be found under any natural conditions. Also, his process requires a mixture of amino acids and nothing else. Only in Fox's laboratory would be found such a specialized mixture; certainly not on the earth, primitive or present.

To round out our criticism, let us add this: The amino acids serine and threonine are essential in body cells, but they were destroyed by the heating in Fox's process.

We have seen that Miller showed that amino acids are easy enough to make; and Fox showed that they can be forced into some sort of linkage. It is appropriate to ask: What are the mathematical odds against *one real protein* forming by chance? These are the odds as calculated by Swiss mathematician, Dr. Charles E. Guye, and quoted by Douglas Dewar in *The Transformist Illusion:*

> If we could imagine unlimited material shaking itself together over vast time so that this material fully interacts, the odds against one protein molecule forming would be 100^{160} (100 multiplied by itself 160 times) to one *against*. That means no chance at all. In fact, to meet those unimaginable odds, there would not be enough material in the whole universe to shake together. We would need more universes of material; and not just three or four more universes; but sextillion, sextillion, sextillion more universes to provide the material.

The *time* required for shaking this material together on

our planet would be, in years, 10^{143}. The mind cannot conceive that sort of time, not even when musing in evolutionary time spans.

Let us just suppose that one lucky protein did beat those infinite odds and did assemble itself; it would be nothing more than an organic molecule. Life would still be a long way off. Other protein substances would have to drop out of the mixture and join this lonely protein.

The question becomes: What are the probabilities of a *complete cell* assembling?

Before attempting an answer, we should consider the magnitude of mathematical expressions. What is the greatest number we can think of? Is it all the atoms in our earth, or in our solar system? Can we imagine the number of atoms, or electrons, in our galaxy; or, yet, in all the galaxies? That is big enough. If the entire universe were solidly packed with electrons it could hold a mere 10^{103} electrons. That seems modest in comparison to the odds we have already quoted, and compared to what we will soon quote.

The evolution theory is built on random chance happenings, and it must involve the laws of chance or probability. They are respectable enough to be used in science, and in business such as insurance. They are used in engineering and architecture to build bridges and skyscrapers; they are used in space exploration. The science of probability was used in developing the quantum model of the atom.

A book, *Evolution: Possible or Impossible?* by James F. Coppedge is discussed in *Bible Science Supplement* (February, 1981). Coppedge quotes the French mathematician, Emile Borel, who summed up the laws of probability in his Single Law of Chance. Borel says that, when the odds are 10^{15} against, then the chance of a given event happening is negligible on the terrestrial scale. When the odds get beyond 10^{50}, there is virtually no chance of it happening, even on the cosmic scale.

Now let us go back to that one lonely protein that beat the odds and which awaits other proteins to join it. The *sec-*

ond protein will be more difficult to assemble because it must match the first biologically. The *third* will be still more difficult. How difficult will be the assemblage of a *full set* of proteins for the simplest cell? According to the above article dealing with Coppedge's book, even after making all possible and even impossible concessions to help the evolutionist's case for generating that first cell, the odds against it happening are estimated as $10^{119,850}$ to one. Whatever may be the margin of error, the mathematics confirm what our reason knows, namely, that spontaneous generation of a living, self-replicating cell is impossible.

Be reassured on this: Science, with its reservoirs of genius and the ultimate in equipment, has not made one speck of life in the laboratory. Publicity surrounded the alleged synthesizing of DNA, of a gene, and of biologically active molecules. These were all tremendous achievements, but they did not create *life.* In every case the experiment began with living cells and manipulated their parts.

Genetic engineering is advancing spectacularly; but this is meddling with existing life, not creating new life.

Dr. Sol Spiegelman of the University of Illinois succeeded in synthesizing biologically active RNA, but he started his process with existing viral RNA and a specific enzyme from a virus. When Spiegelman was asked if he had created life in a test tube, he replied humbly: "Only God can create life."

In 1970, one of the most eminent men in science, Sir Ernst Chain, F.R.S., who, for his work on penicillin, has received a Nobel Prize jointly with Fleming and Florey, said: "The laboratory synthesis of even the simplest cell is just not on; and the notion that man is about to compete with God is absurd . . ."

Such men acknowledge that there is a dimension which science cannot reach.

A recent news item announced that two prominent men have changed their views and now deem that a deity is necessary to explain first life. Erstwhile agnostic, Sir Fred Hoyle has long been a world figure in science. Erstwhile atheist, Chandra Wickramasinghe, Professor of Applied Mathematics and Astronomy at University College, Cardiff, has sometimes collaborated with Sir Fred.

The *Sunday Mail*, Brisbane (September 20, 1981), in a feature article said that they are "changed men. They are both believers."

The article explained:

> What convinced both men were calculations they each did independently into the mathematical chances of life starting spontaneously. When each had finished, they looked at the answer almost in disbelief. Each found that the odds against the spark of life igniting accidentally on earth were staggering—in mathematical jargon "10 to the power of 40,000."

Whether they used a different formula, or different specifications for what constitutes life, their figure, $10^{40,000}$, is not as spectacular as a figure I quoted earlier, but it has an equally emphatic message: "Impossible!"

Hoyle seems to have accepted the changed view more easily than Wickramasinghe did. Wickramasinghe is quoted as saying:

> I am quite uncomfortable in the situation, the state of mind I now find myself in. But there is no logical way out of it . . . We used to have open minds; now we realize that the only logical answer to life is creation—and not accidental random shuffling.

We must admire the moral fiber of both men publicly making such an honest admission in today's intellectual climate. Still, we must not overstate the case. It seems they continue to believe in evolution, not by Darwinian gradual steps, but an evolution (on Earth) by a series of leaps; and that this was begun by life spores arriving on earth from space—life spores that were not created by accident.

As to the question: What is God? the article reported that they suggest that God is the universe.

Christians have firm ideas about a personal Creator God. It is fortifying to find that science and mathematics, far from dispensing with the Creator, point straight to Him.

Blind, mindless chance could not assemble the first living cell; nor could it evolve trillions of cells into the fantastic mechanism of my human body. Incontrovertibly I can say that I am wonderfully and supernaturally made.

In a relatively lesser sense, the meanest microbe can make a similar claim.

REFERENCES

1. Dr. Duane Gish; Articles in *I.C.R. Impact* Series (Nos. 31, 33, 37); also in *Bible Science Newsletter,* September, 1977.
2. Dr. C. B. Gower; article in E. P. M. Pamphlet No. 220 (October, 1978).
3. Dr. Wayne Frair, Dr. Harold Armstrong, Wilbert Rusch, Sr.: articles in *Why Not Creation?*
4. *The Creation Explanation* (p. 96 *et seq.*).
5. *Scientific Creationism* (pp. 59-65).
6. *The Transformist Illusion* (pp. 9-11).

Chapter 10

THE UNIVERSE AND THE EARTH

"In the beginning God created heaven and earth."

Those opening words of *Genesis* have been widely replaced by: "In the beginning was the Big Bang." The hypothesis is that all matter/energy was originally concentrated in a super-dense state; that it exploded, and the outrushing material eventually formed into hydrogen and some helium. Then, after great time, those outrushing gases arranged themselves into stars and galaxies and planets and animals and man.

We object that such complex workmanship by simple hydrogen would make a mockery of the 2nd Law of Thermodynamics. Furthermore, there is no explanation of where the super-dense matter/energy came from. Also, there is the objection that, by classical physics (non-relativistic physics), the super-dense concentration of matter would not explode, but would collapse under its own gravitation into a great "black hole."

Even if we suppose that it did explode, it would simply expand as a homogeneous cloud. It could not construct the marvelous order of the universe and of life. In the words of Sir Fred Hoyle:

> Even though outward speeds are maintained in a free explosion, internal motions are not. Internal motions die away adiabatically and the expanding system becomes inert, which is exactly why the Big Bang cosmologies lead to a universe that is dead-

and-done-with almost from the beginning . . . The notion that galaxies form, to be followed by an active astronomical history, is an illusion. Nothing forms. The thing is as dead as a doornail. (Hoyle, 1981. *New Scientist,* 92:521-527.)

The detection of a predicted cosmic background radiation was claimed to confirm the Big Bang. However, J. Narliker suggests that the microwave radiation is just as likely of astrophysical origin and not a relic of the Big Bang. (Narliker, 1981. *New Scientist,* 91:19-21.)

There are so many serious objections to the Big Bang, so much "rigging" of conditions to make it work, that scientists admit that the Big Bang cannot be explained naturally. It has to be explained as a "singularity," in which the physical laws (as we know them) did not apply. We ask: How can a "singularity" be justified as science?

The Big Bang is a materialist device to avoid creation. The new science cannot tolerate the concept of stars and galaxies having been *created* in their present form. Stars and galaxies are required to *evolve* from out-rushing clouds of gas—something that just could not happen.

Stars: A vast cloud of diffuse gas would not condense into stars. By natural laws the gas would expand and become more diffuse.

There is no scientific justification for describing some stars as "young," some as "mature" and some as "old." A star supposedly burns at a rate dependent on the cube of its mass. Theoretically, starting with a large mass, the star burns fast. Then, as its mass reduces, it burns more slowly. If stars really have been forming and evolving during billions of years, we should now observe a range of differences in the chemical composition of various stars. Instead, we observe the same chemical composition in them all, in the "old" and the "young," and even in the material between the stars.

O Stars and B Stars: These huge "young" stars are supposedly burning themselves so fiercely that, if we project the process backwards, just a few thousand years ago they would have had a nearly infinite mass—which is impossible.

Binary Stars: These are pairs of stars circling each other. Often one star is classed as "young" and the other as "old," though they are supposed to have formed together. The most beautiful star, Sirius, is a pair. Sirius A is bright, supposedly burning fuel at a prodigious rate and therefore "very young." Sirius B is a white dwarf, which means it has completed its "evolution" through all star stages to the end of the line, the oldest of the "old." How can evolution explain the old and the young in Sirius and the many other odd couples in binary stars? To suggest that one star must have evolved faster than the other is to abandon science and enter the realm of "we'll think of something."

Galaxies: *Formation* of galaxies from primordial matter is a total enigma. A further enigma is this: How do galaxies *preserve* their shapes, spirals and ellipticals, through long ages? *The Encyclopedia Britannica* (1964) called this problem area "a challenge to cosmological thought."

Galaxy Clusters: Galaxies are always found in clusters; never alone. The clusters are held together by the force of their own gravity. The problem is that individual galaxies have random motion; and all clusters have insufficient gravity force to hold the cluster together. Consequently, after a short period of time, many individual galaxies should have escaped from their cluster, and should now be observed as lone galaxies. But no lone galaxies are observed. After some millions of years, all galaxies should have escaped. There should not be any clusters of galaxies now remaining.

Spiral Galaxies: These rotate faster at the center than further along the arms. After only one or two revolutions, the spiral arms should be wound up tightly. Some do have the appearance of being wound up into a huge rotating center. However, what do we make of the large number of spiral galaxies (including our own Milky Way) with arms still trailing beautifully, but supposed to be billions of years old?

Barred Spiral Galaxies present an extra impossibility. Evolutionists cannot explain why the straight bar, like an axle at the center, is still straight. The bars should start to bend at the beginning of the first revolution, long before the spiral arms begin to wrap up. How could they remain straight after billions of years?

 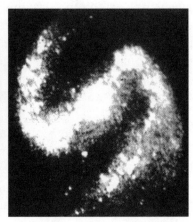

Spiral Galaxy   *Barred Spiral Galaxy*

Red Shift and an Expanding Universe: Astronomers noticed that the spectral light from distant galaxies is "shifted" toward the red end of the spectrum. They interpreted this on the Doppler principle. They deduced that the galaxies must be speeding away from us, thereby stretching the light wavelengths, and making the spectral light seem redder. Galaxies at greater and greater distances had greater and greater red shifts. This suggested that the further away a galaxy is, the greater is its speed. Soon, the red shift became a measure of a galaxy's distance and its speed.

In 1929, astronomer Edwin Hubble put forward the theory that, as galaxies are speeding away from the observer in every direction, the universe must be expanding. The expanding universe was assumed to be the result of the Big Bang explosion 15 or 20 billion years ago (estimates vary).

The red shift is real, but the Doppler interpretation is assumption. Other interpretations are being proposed for the red shift.

Astronomer Dr. Halton Arp of the Hale Observatories in Pasadena, California seriously questions the doppler interpretation. He presents evidence that many galaxies, and most quasars, have discordant red shifts. This means that their red shifts do not correspond to their theoretic distances.

Even more damaging is Dr. Arp's evidence that some bodies, which are grouped at about the same distance from us, have different red shifts. (Ref. *Bible-Science Newsletter,* August, 1974, p. 2.)

Quasars present special problems. They are radiating much more energy than their size would permit. This indicates that they cannot be as distant as their red shifts denote. Furthermore, a quasar can have more than one red shift. One has been observed to have as many as five red shifts. (Ref. *The Creation Explanation,* p. 153.)

A recent querying of the Doppler interpretation was reported in *The New Scientist* (June 20, 1985) in an article by John Gribbon titled "Galaxy Red Shifts Come in Clumps." Gribbon says that any suggestion that the simple interpretation of the red shift may be wrong, or at least incomplete, sends shivers down the spines of conventional cosmologists. His article then proceeds to discuss new theories of the red shift by W. G. Tifft of the University of Arizona.

Tifft proposes that the red shift is somehow an intrinsic property of a galaxy—as intrinsic property that is not related to expansion velocity at all; an intrinsic property whereby red shifts take up preferred states, like the quantized energy within an atom.

Well, that must send shivers down some spines; but Gribbon says that whether Tifft is right or wrong, his work is now respectable enough to give both observers and theorists something to chew on.

To sum up: The hypothesis of an expanding universe and a Big Bang depends on the Doppler interpretation of the red shift. If it is wrong, modern cosmology falls apart. So much controversy surrounds the Doppler interpretation that we may disbelieve the expanding universe, and its Big Bang, and still be in good scientific company.

Has Light Slowed?

If asked to name one thing that is ever constant, most people would vote for the speed of light, the unchanging constant "C," on which many other physical constants depend, directly or inversely. However, about 1981, a controversial hypothesis has claimed that light has been slowing exponentially.

Barry Setterfield, an Australian whose fields are geology, physics and astronomy, had been examining the many measurements of light speed during 300 years. He found they revealed, not a constant speed as he had been taught, but a progressive slowing according to a precise geometrical decay function. He was assisted by Trevor Norman, tutor in Mathematics at Flinders University, South Australia. The raw data was fed into the University's computer. The result indicated that light speed has been decaying exponentially. The decay curve that clearly was the best fit for all the data was selected. Extrapolating backwards the curve indicated that the speed of light 6,000 years ago would have been close to infinity, which could be regarded as the beginning point, and which (without any nudging by Setterfield) coincided reasonably with the date of Creation computed from Biblical chronologies.

Setterfield had come up with a radical proposition. If true, it would upset some things which modern science accepts as basics, e.g., radioactive decay rates; red shifts; cosmic ages. On the other hand, it would provide some answers which creationists are seeking. It does not disturb the structure of basic science; but it would upset things basic to evolutionism. Not surprisingly, it has been widely denigrated and ignored. Even some leading creation scientists avoided it.

Setterfield kept researching as objections were raised. He has made an important adjustment to a point which had been causing criticism. Originally his thesis showed that the decay ceased in the year 1960, and from that date the speed remained constant. He has recently found that the post-1960 measurements indicated constant speed because of a technical mistake in method of timing. Having corrected this error, he finds that the decay continues after 1960. Thus he removed what was an objection against his work.

A 1986 book, *It's a Young World After All,* by Paul D. Ackerman, Ph.D., entertainingly examines many evidences for a young earth, including Setterfield's work. Ackerman states (p. 77): "To date no one has been able to debunk the findings, and corroborations of Setterfield's work seem to be piling up."

Evolution-leaning scientists have been hostile. Creation scientists have been divided—some for; some against. Latest information from the Creation Science Foundation, Brisbane, Australia, is that some creation scientists who were wary of Setterfield's hypothesis have now begun to accept it.

Polonium Radiohalos
("The Creator's Signature")

When a radioactive element decays in a transparent mineral (such as in mica in granite rock), the emitted alpha particles infuse a spherical shell of discoloration in the mineral.

When uranium decays it imprints its characteristic sphere

of discolor (its radiohalo) in the rock. Also, the uranium decay chain produces three isotopes of Polonium. These, in turn, imprint in the rock their respective halos of discoloration.

We are most interested in the isotope Polonium 214. It is born in the uranium chain, and then itself decays very rapidly. It disintegrates and is gone in less than a minute. In its short life, Polonium 214 imprints in the rock its own identifiable radiohalo.

Today, Polonium is never found except as a product of uranium decay. However, in the Pre-Cambrian rocks there are trillions of mysterious Polonium radiohalos. The mystery is that the Polonium particles which formed these halos were never associated with a uranium parent. Every test has eliminated the possibility of a uranium parent. They must have been free particles of the element Polonium. They must have come into existence as free and independent elements, lived their few moments of life while they infused their halos in the surrounding solid rock, and then disintegrated and vanished.

Radiohalos will not form in molten rock, nor in very hot rocks. They form only in solid rocks that are cooler than 300°C.

Hypothetical "Big Bang" rocks would be molten, and would require long ages to cool and solidify. Long before that happened, any free Polonium would have disintegrated and disappeared. Polonium radiohalos knock out the idea that the Pre-Cambrian rocks ever existed in molten state. In supposed "Big Bang" rocks there could be Polonium radiohalos only from Polonium produced by uranium after the rocks had cooled and solidified.

Evolutionists have no solution to explain the trillions of radiohalos that were imprinted by free Polonium 214. They acknowledge that this is a mystery.

Creationists have no problem. Their answer is that the primordial Pre-Cambrian rocks came into existence suddenly, instantaneously. At the same instant, free primordial Polonium came into existence inside those solid *new* rocks. The Polonium immediately imprinted its halos; and, after a few moments, the free Polonium was gone, while the solid rocks (marked with the halos) remained. In other words, in an instant of creation, solid basement rocks and free Polonium appeared; the Polonium quickly disappeared; and the rocks remained bearing the message of the radiohalos.

The international authority on radiohalos is the highly respected Dr. Robert Gentry. He acknowledges that the study of radiohalos forced him to regard them as evidence of instantaneous creation of the earth. Polonium radiohalos changed Dr. Gentry into a believer in literal six-day creation. (Ref. *Bible-Science Newsletter,* January, 1982, October, 1984; Also, *C.R.S. Quarterly,* December, 1984. Cf. also Robert V. Gentry, *Creation's Tiny Mystery*, Earth Science Associates, Knoxville, TN, 1992.)

The Solar System

The solar system defies explanation. For 300 years there has been a succession of theories on its origin—all of them unsatisfactory. The latest theory is by Gerald Kuiper; but he acknowledges its weaknesses. He goes further and says that perhaps the problem of the solar system has no scientific solution. The most popular theoretic theme is that a rotating cloud of gas and dust condensed and formed the sun and its planets. Let's note some insuperable problems:

- The cloud would not condense; force of gravity would be insufficient. Gravitational attraction would not be effective until the particles were the size of the moon.
- The particles would not accumulate to form planets. Science knows of no process whereby grains of dust will stick together and accumulate to a size where gravity will take effect.

- The sun is rotating much too slowly to satisfy the theory. The sun's mass is more than 99% of the total mass of the solar system. If it had condensed from a cloud, the sun should have 99% of the rotational momentum of the solar system. Instead, we find that the lazily turning sun has less than 1% of the rotational momentum, while the insignificant planets have more than 99%. How could the sun have transferred almost all of its rotational momentum to the tiny planets? Geniuses have tried to answer that; and still they try.
- All nine planets orbit the sun in the correct direction, and seven of them rotate correctly on their axes. But two of them, Venus and Uranus, rotate backwards. As well, Uranus lies flat on its back with its axis almost along the plane of orbit around the sun.

  —Of the 31 moons in the solar system, 20 orbit their planet in the same direction as the planet rotates on its axis; but 11 orbit their planet in the reverse direction.

  —The planet Pluto orbits the sun on a plane different from that of the other planets.
- Space exploration has shown that our moon is very different from the earth. Its origin is a total mystery. It could not have been torn from the earth; nor could it have formed from the same cloud as the earth.
- In the sun and in the universe, hydrogen and helium are abundant, but the heavier elements are almost non-existent. In the inner planets, Mercury, Venus, Earth and Mars, the heavier elements are as abundant as hydrogen and helium, which makes these planets curiosities in the universe. It is difficult to see how they could have formed from the same cloud as the sun allegedly did.

With these and many other problems, the solar system bristles with difficulties for the theorists. Science could save valuable time by facing up to the inevitable answer, as Isaac Newton did, namely, that the solar system is the product of intelligence, not of chance.

(Ref. *The Origin of the Solar System,* by J. C. Whitcomb; *Why Not Creation,* article by George Mulfinger, pp. 39-66; *The Creation Explanation,* pp. 140-143.)

Young Earth: Young Solar System: Young Universe

There comes the question: Is the earth old or young? What of the universe?

The popular billion-year ages have grown out of two ideas: the assumption that evolution happened, and radiometric dating. If these two are found wanting, then so are the billion-year ages found wanting.

On the other hand, there are many natural phenomena which seem to limit earth and universe to an age of only some thousands of years. We look at a few of them.

Carbon-14 Imbalance: For equilibrium to be achieved between production and decay of Carbon-14, only about 30,000 years would be required. However, as previously discussed, equilibrium is far from achieved. Production in the upper atmosphere exceeds decay on earth by 30-40%. In the absence of a better explanation, this means that the cosmic rays have been acting for less than 30,000 years. Melvin Cook and Robert Whitelaw, working on different equations of non-equilibrium, estimated the age of the atmosphere at 10,000 years and 5,000 years respectively.

Helium: When uranium decays, it produces not only lead, but also helium. The helium passes into the atmosphere. The puzzle for evolutionists is that there is not nearly enough helium in the atmosphere. If uranium has been decaying for billions of years, there should be a lot more helium. The evolutionists tackle the problem by an assumption that the missing helium must simply have escaped into space. But there are several studies which show that helium is probably *entering* the atmosphere from the sun's corona (e.g., Prof. Melvin Cook in *Nature,* 26/1/57). And there are studies that show that the proposed escape of helium in sufficient quantity to meet the problem is unlikely. This is covered by Larry Vardiman, Ph.D., in *The Helium Escape Problem* (I.C.R. *Impact,* May, 1985). Vardiman goes along with the view that the escape problem is one that "will not go away and is unsolved."

Creationists contend that the helium in the atmosphere is mostly original primordial helium, together with some helium accumulating from radioactive decay during some thousands of years.

Comets: Short-period comets complete their orbits in less than 200 years. Every time a comet circles the sun it has a moment of glory as the beautiful tail forms. But it pays for this glory by losing part of itself, by losing mass at a prodigious rate. Comets are relatively small, averaging perhaps a kilometer in diameter. They cannot sustain many of these moments of glory before they disintegrate.

Astronomer R. A. Lyttleton estimated that no short-period comet could survive for more than 10,000 years (*Mysteries of the Solar System,* 1968). Recent assessments of Halley's Comet may allot to that big comet a much longer life than 10,000 years, but it must still be a very limited lifespan.

Multitudes of surviving short-period comets are heavenly witnesses that the solar system is very young. They are a problem for evolutionists who require a multi-billion year old solar system. They meet the problem by proposing that there must be a huge number of spare comets waiting out in space; and they suggest that, from this postulated reservoir, new comets are kicked into the solar system to replace old comets. There is no evidence to support the hypothesized cloud of spare comets, the Oort Cloud, as it is called. It actually falls apart under critical scientific analysis. Furthermore, short-period comets have parabolic orbits, not hyperbolic orbits, and this is proof that they were not kicked from space into the solar system. All this is convincingly covered by Paul Steidl in *Bible-Science Newsletter,* May, 1986, pp. 12, 14.

Short-period comets are vivid evidence that the solar system is very young.

Earth's Magnetic Field: Dr. T. G. Barnes, Professor of Physics at the University of Texas in El Paso, has studied and updated previous scientists' work on the strength of the earth's magnetic field, which has been carefully measured during nearly a century and a half. Its strength is decaying exponentially with a probable half-life of 1,400 years. This means it was twice as strong 1,400 years ago; four times as strong 2,800 years ago. Projecting forward, it means that there will be no magnetic field remaining in the year 4,000 A.D.

Kofahl and Segraves discuss these findings in *The Creation Explanation* (p. 194). They say that, if the present magnetic decay had been going on for 30,000 years, the decaying energy, being transformed into heat, would have generated enough heat to have completely vaporized the earth. They say that this evidence supports an earth history of not much more than 10,000 years.

Prof. Barnes himself says, "It would appear from these arguments that the origin of the earth's magnetic moment is much less than 20,000 years ago." (*Speak to the Earth,* p. 309)

Oil and Gas Wells: That oil and gas are still at extremely high pressures is something of a mystery for modern geology. If oil and gas had been trapped in the rocks for millions of years, the pressure would have seeped through the rocks. No excess pressure should remain at all if the formations are more than some thousands of years old. (*The Creation Explanation,* pp. 192-193.)

The Poynting-Robertson Effect: This means that dust and stones orbiting the sun are being struck by the sun's radiation. This decelerates them, and they gradually spiral into the sun. The smaller the particle, the faster it is swept into the sun. This effect should have cleared our solar system of dust, and of particles up to 4 c.m. diameter, in less than 200 million years. This is a small fraction of the solar system's supposed age of 4-1/2 billion years; and yet, the dust is still there.

Then, if we look out to the really big suns (the O stars and the B stars) with 100,000 times the radiation of our sun, those giants should sweep up their dust 100,000 times faster than our sun does. Yet, these giant stars are nearly all surrounded by huge clouds of dust and gas. Here is evidence of a young universe.

There is another problem with those O and B stars. They are burning up their energy at a prodigious rate. To support the energy consumption each of them would need to have been an "infinite" mass not many thousands of years ago. This is realized by evolutionists. The simple answer is that the universe is only some thousands of years old. These and other arguments are well presented by Harold Slusher, geophysicist and astrophysicist, in "A Young Universe," in *Bible-Science Newsletter* (January, 1975).

Meteorite Dust: This fine dust is constantly falling through the earth's atmosphere. In 5 billion years it should have formed a layer about 130 feet thick if accumulating undisturbed. There is no sign of such a layer. Meteorite dust is rich in nickel, but in earth rocks, nickel is rare. In billions of years the nickel-rich dust should have made our oceans rich in nickel. In fact, ocean water and ocean sediments have so little nickel that the meteorite dust could have been falling for some thousands of years at most.

The moon tells a similar story. Meteorite dust has been falling on the moon's surface just as on earth. On the moon there is no wind, water or weather to disturb the dust as it accumulates. Assuming an age of billions of years for the moon, it was feared that the astronauts' landing craft might sink into 60 or 100 feet of dust, so the landing craft was fitted with pancake landing feet. When the astronauts landed on the moon they found half an inch of dust, which indicated that the moon has been there for about 8,000 years. (Ref. *Bible-Science Newsletter,* March, 1975 and Supplement April, 1979; also *The Creation Explanation,* p. 190.)

How Do Stars Burn?

The question of what makes the sun burn has long been pondered.

Chemical Burning is inadequate. It could last only about 5,000 years.

Gravitational Collapse: Around 1850, Hermann von Helmholtz proposed that the sun's incandescence comes from the sun's gravitational shrinkage. He calculated a burning life of about 20 million years. His theory was rejected because, and only because, evolution requires billions of years.

Nuclear Fusion: About the beginning of this century came knowledge of radioactivity. Nuclear fusion could be the source of the sun's energy. Evolutionists grabbed it because nuclear fusion in sun and stars could go on for billions of years. It became accepted dogma, even though it has no better scientific justification than gravitational shrinkage.

Is it correct? If nuclear fusion provides the sun's energy, the sun should be showering earth with neutrino particles. Ingenious neutrino traps have been counting neutrinos for more than a decade; but not nearly enough neutrinos have been detected to support nuclear fusion in the sun.

Failure to detect solar neutrinos is a severe upset to the basic assumption of astrophysics, viz., that nuclear fusion makes the sun (and other stars) burn. Scientists should now reconsider the alternative of **Gravitational Collapse**, especially because there is direct evidence that the sun is shrinking.

In 1979, in a paper presented to the American Astronomical Society, solar physicist John Eddy and Mathematician Aram Boornazian said that measurements of the sun's diameter made at Royal Greenwich Observatory from 1836 to 1953 indicated shrinkage of 2 seconds of arc per century. At this rate the sun would disappear within 100,000 years. This spurred other research which found that the Greenwich records contained a consistent error. Then in 1980, two groups of scientists made separate findings. Their figures were in agreement, but one team claimed that the sun is shrinking, the other that the sun is constant.

The breakthrough came in 1981. One of Eddy's colleagues, Ronald Gilliland, using the Greenwich records (corrected for error) and four other sets of data, found that the overall shrinkage of about 0.1 seconds of arc per century since the early 1700's is real. He used cautious words in his statements: "Given the many problems with the data sets, one is not inexorably led to the conclusion that a negative secular solar radius trend has existed since A.D. 1700, but the preponderance of current evidence indicates that such is likely to be the case."

He also found a minor pulsation of 0.02% of the radius with a rhythm of 76 years, and a further tinier pulsation which ties in with the 11-year sunspot cycle. That overall shrinkage would mean that the sun was twice as large two million years ago. That posed two alternatives: either the shrinkage must be explained away, or a young age for the sun must be accepted. Evolutionists have adopted the former alternative. They suggest that, as small pulsations of 11 and 76 years have been identified, the overall shrinkage may be interpreted as a phase in a large cycle of contraction and expansion yielding long-term stability. This suggestion is not verifiable except by future data observed over hundreds of years.

Meanwhile the observed facts are:

(a) Solar neutrinos are missing, which should discredit the assumption that the sun (and other stars) are burning by nuclear fusion; and

(b) Direct evidence that the sun's diameter has shrunk during three centuries is prima facie evidence that gravitational collapse is a tenable theory for the burning energy of sun and stars. However, that would imply that the total life of the sun be limited to about 20 million years, a minute fraction of the billions of years demanded for evolution.

(Ref. *New Scientist,* March 3, 1983, pp. 592-595. Also articles by Hilton Hinderliter and Paul Steidl, *C.R.S. Quarterly,* June, 1980.)

Now here is something interesting:

Three centuries ago James Ussher, Archbishop of Armagh, using Biblical chronology, estimated that creation took place in the year 4004 B.C. Our scientific age has had many a long laugh about the Ussher estimate. But now the laughing should stop.

In 1978, solar physicist Dr. John Eddy, speaking at Louisiana State University, made an admirably honest statement. Eddy, an evolutionist, acknowledged that he believes the sun is 4-1/2 billion years old; but he went on to qualify this. He said: "However . . . I suspect that we could live with Bishop Ussher's value for the age of the earth and sun. I don't think we have much in the way of observational evidence in astronomy to conflict with that." (*Geotimes,* 23(9):18-20).

The eminent solar physicist's words must mean that the Bishop's estimated age of 6,000 years for earth and sun is every bit as worthy of belief as the modern cosmologist's estimate of billions of years.

Chapter 11

OUT OF NOTHING

The moon is lifeless. In earth's embrace, it reminds us that space is hostile to life. We search the solar system and find that every foothold is hostile to life. Within our knowledge, there is no hospitable place in space; only a universe at enmity with life. In this sterile immensity the earth is an alien. Bursting with life, our earth is unique—the one freak spot designed for life.

Today there is a mounting mania for supposing other earths and other life out there in all the corners of the universe. In this modern thinking, the starting point is to exclude anything supernatural. Creation is unthinkable. Therefore, life must have begun from dead matter by natural chemistry, by spontaneous generation. That is the bad science which the great Louis Pasteur confronted long ago.

When Pasteur had completed his remarkable experiments, he had scientifically annihilated the old idea of spontaneous generation of life. He showed that only life can beget life. Since Pasteur's day something new has gotten into science, something called godlessness. Science rejects the supernatural. Science is now forced to go back to pre-Pasteur darkness and say that life exists, but we must not admit a Creator. So we have to say that matter spontaneously burst into life—despite Pasteur. Let's have spontaneous generation just once, way back in the mysterious past, despite the laws of thermodynamics and mathematics and common sense. This sort of science would rather break the laws of science than admit a Creator.

Earth

Moon

The scientist who rules out Creation will find it "logical" to say that, if one solar system has come into existence by chance, then many other solar systems must also have come into existence in the same way—out there in space. If living matter happens to exist on earth, life must be a natural happening; so there must be other life on those other earths that we imagine out there.

Whether or not there exist other solar systems (planets revolving around a sun), we have to admit that even our own solar system is inexplicable. It is a mystery. No scientist can explain it. Theories have come and gone for 300 years. Isaac Newton was critical of one of the earliest theories. Newton said that no natural cause could have produced the solar system. He said the solar system must have been the effect of counsel, that is, of intelligence. The position today is this: After 300 years of one theory after another, we are no nearer than we ever were to explaining the solar system. This is admitted.

Undaunted by the unfathomable solar system, the theorists render their accounts explaining the entire universe. Currently popular is the "Big Bang," which allegedly exploded an immense mass of primordial matter and sent it rushing outwards, whereupon, in full flight, it assembled itself into stars, and galaxies of stars, and into the wonderful universe. Prideful man, with scraps of data but impatient for answers, is building hypotheses upon hypotheses and is leaving out the key of it all, the Creator of it all.

Before explaining the material universe, its birth and formation, it would be wise to understand what matter itself is, elusive matter, dissolving into the immaterial when probed. And who can decipher matter's other self, energy? Who can pin down light? Who can unravel gravity?

Everywhere we are faced with the inscrutable, while godless theories try to explain the inexplicable, try to explain the watch without the watchmaker.

If the universe was created, then the universe is a pure act of God; and you can't explain that by uniformitarian theory.

Apparent Age: There would have been a moment after Creation when apparent age was different from real age. Note well the point I am trying to make—apparent age versus real age.

Think of the miracle at Cana when water was changed into wine. A hard-headed wine-taster would taste the wine and say: Old wine! And he would be wrong. The wine's apparent age was years; its real age, a few minutes.

Remember the miracles of Lourdes. Take a cure medically certified as beyond the natural: A man whose leg was broken and the bone decayed, so that there was a gap of missing bone. In the instant of the miracle, new bone filled the gap. An impersonal specialist might examine the leg and say: This bone is 30 years old; and he would have been wrong. The apparent age was 30 years; the real age a few minutes.

So, our no-nonsense scientist observes the magnificent universe and he tries to explain *it* all the way back to some pre-elemental soup.

Unless we are completely atheist we accept a Deity—a God. If a Deity exists then that Deity, with His "Fiat!"—

"Let it be made!"—could have brought forth instantly a perfect universe from nothingness. Even Huxley conceded this.

Thomas Huxley, the agnostic, Darwin's greatest ally, "Darwin's bulldog"; even Thomas Huxley, archbishop of evolution, hater of religion; even Huxley conceded that, given a Deity, he would have no difficulty in conceiving that, where nothing had existed, the universe could suddenly have appeared out of nothingness, at the volition of that Deity.

I now suggest to you, and this suggestion is supported by the faith of many hundreds and perhaps thousands of highly qualified scientists—I suggest that the Creator's "Fiat" produced a fully-fledged, perfectly operating, adult universe. And, from the dust of the earth, another "Fiat" produced an adult man. And, when that man looked up, I suggest he would have seen God's handiwork in the stars of the sky without waiting years for light to cross space.

At that moment the apparent age of the universe might have seemed billions of years to a uniformitarian scientist, but its real age was maybe six days. Adam's apparent age would have been, say 21 years; but his real age, a few minutes, on that newly created earth.

Suppose we could have cut down a tree in the Garden of Eden. It is possible that it would have shown the usual rings of growth that are inherent in the tree; its apparent age would thus have been many years, but its real age a few days. If a geologist could have entered the Garden of Eden and tested a rock sample, it is quite likely he would have found some potassium and some argon and he would have dated the rock at millions of years—the newly created rock!

If the earth and the universe were created, then we must venture where science cannot follow, where Christian scientists stop "explaining" and start adoring.

This proposal of apparent age versus real age is not some cute idea that you can take or leave. Unless you accept it, you cannot accept any original creation. On the other hand, regardless of what God may have chosen to create, the perverse mind of man can invent some sort of evolutionary history to precede it, to "explain" it. But to do the explaining, man bends and breaks the fundamental laws of nature. He breaks the unbreakable laws of thermodynamics at every twist and turn as he explains the watch without the watch-Maker.

Today, when the theory of evolution can be shown to be not credible, we behold the paradox of a new surge of evolutionary propaganda flooding the world through the mass media and our educational systems. It must be terribly important to some people to persuade men that they are only animals and that science needs no God.

If you are wondering why, the following will help to explain. Newman Watts, a London journalist, wrote a book entitled *Britain without God.* In his research for that book he discovered something. He discovered that those who would shoot Christianity to pieces are using the bullets of evolution. This is his warning, decisive and clear:

"Every attack on the Christian Faith made today has, as its basis, the doctrine of evolution."

Appendix A

THE CHURCH'S POSITION

It is fundamental that we believe in Creation, out of nothing, of Heaven and earth by one almighty personal God whose power now sustains His creation. (Fourth Lateran and First Vatican Councils).

We may believe in evolution of the body (if convinced of it on the evidence),* but not evolution of the soul; and regarding the origin of the earthly race of man, polygenism (the idea that there were many Adams) is forbidden. (*Humani Generis*).

Any idea of a god evolving with the universe was condemned by the First Vatican Council.

In the 20th Century, with the growth of evolution ideas, Pope Pius XII made clear the Church's position:

Firstly, in 1941, in an *Address to the Pontifical Academy of Sciences,* the Pope said that *Genesis* attested these certainties, with no possible allegorical interpretation:
   (1) Man's essential superiority to other animals because of his spiritual soul.

---

*More precisely, in *Humani Generis* Pope Pius XII said that "research and discussions" regarding evolution of the human body may take place by men *experienced in both science and theology*. (See pp. 169-171 herein.) The Pope referred to "the present state" (1950) of the human sciences; since that time, science has more and more shown the theory of evolution to be untenable. —*Publisher*, 2000.

(2) In some way the first woman was derived from the first man.

(3) The first man could not have been generated literally by a brute beast in the proper sense of the term, without divine intervention.

Secondly, in 1950 Pope Pius XII issued the encyclical *Humani Generis,* which dealt with various modern errors. He pointed out how evolutionism can lead to serious error:

> A glance at the world outside the Christian Fold will familiarize us, easily enough, with the false directions which the thought of the learned often takes. Some will contend that the theory of evolution, as it is called—a theory which has not been proved beyond contradiction even in the sphere of natural science—applies to the origin of all things whatsoever. Accepting it without caution, without reservation, they boldly give rise to monistic or pantheistic speculations which represent the whole universe as left at the mercy of a continual process of evolution. Such speculations are eagerly welcomed by the Communists, who find in them a powerful weapon for defending and popularizing their system of dialectical materialism; the whole area of God is thus to be eradicated from men's minds.
>
> These false evolutionary notions, with their denial of all that is absolute, or fixed or abiding in human experience, have paved the way for a new philosophy of error . . . (Pars. 5-6).

He referred to reliance on "the positive sciences" and said that this is "praiseworthy" when they deal with "clearly proved facts"; but we must be cautious when they are "hypotheses, having some sort of scientific foundation," which involve Church doctrines. He continues, and applies this to *man's body and soul:*

> For these reasons the Teaching Authority of the Church does not forbid that, in conformity with the present state of human sciences and sacred theology, research and discussions, on the part of men experienced in both fields, take place with regard to the doctrine of evolution, in as far as it inquires into the origin of the human body as coming from pre-existent and living matter—for the Catholic faith obliges us to hold that souls are immediately created by God. However this must be done in such a way that the reasons for both opinions, that is, those favorable and those unfavorable to evolution, be weighed and judged with the necessary seriousness, moderation and measure, and provided that all are prepared to submit to the judgment of the Church . . . (Par. 36).

He then deplores the rashness of those who abuse this liberty of debate by treating evolution of the body as if proved beyond doubt.

Next he moves to *Polygenism:*

> Christians cannot lend their support to a theory which involves the existence, after Adam's time, of some earthly race of men, truly so called, who were not descended ultimately from him, or else supposes that Adam was the name given to some group of our primordial ancestors. It does not appear how such views can be reconciled with the doctrine of Original Sin, as this is guaranteed to us by Scripture and Tradition, and proposed to us by the Church. Original Sin is the result of a sin committed, in actual historical fact, by an indi-

vidual Adam, and it is a quality native to all of us, only because it has been handed down by descent from him. (Par. 37). (A footnote reference to *Romans* 5:12-19, and Council of Trent, session V, can. 1-4, indicates that this is well established Church teaching.)

Note: Father McKee (in *The Enemy within the Gate*) summarizes that the clear intention of the encyclical is to exclude polygenism from theology. He adds that this part of the encyclical teaches that Adam was an individual man, not a group, and his sin was an actual historical sin which is passed on to us by blood descent.

Further note: *Humani Generis* expressly states that, in encyclicals, a Pope is teaching as Vicar of Christ, clarifying what the Church already teaches, and this removes the subject from free debate among theologians. Despite this, many theologians still strive to outflank *Humani Generis* in efforts to reconcile Original Sin with polygenism.

"Mystici Corporis" (1953): Pius XII reinforced *Humani Generis* with this encyclical. Part of its teaching is summarized by Father McKee: It includes (1) Adam was the father of the whole human race; (2) he was created in perfection; (3) all mankind inherited the stain of his sin.

Address by Pope Paul VI (1966): Paul VI addressed a group of theologians and reminded them that "Catholic doctrine on original sin was reaffirmed in the Second Vatican Council" (in *Lumen Gentium* and in *Gaudium et Spes*) "in full consonance with divine revelation and the teaching of preceding Councils of Carthage, Orange and Trent." (Ref. *Lumen Gentium* section 2, *Gaudium et Spes* sections 18, 22 and 24.)

He reproved some modern authors whose explanations of Original Sin seem "irreconcilable with true Catholic doctrine." He affirmed Church teaching "according to which the sin of the first man is transmitted to all his descendants, not through imitation but through propagation" (i.e., through human descent).

He also reaffirmed the special creation of each human soul by God.

The Catholic Catechism (Fr. J. A. Hardon, S.J.) states on page 106:

> While never formally defined, the fact of a direct creation of each individual soul belongs to the deposit of the Christian faith. Implicitly taught by the Fifth Lateran Council . . . it is part of that vast treasury of revealed truths which are jealously safeguarded by the Church. This was brought to the surface in *Humani Generis,* in 1950 . . .

"Credo of the People of God" (1968): Pope Paul VI again clarified the Church's teaching that our first parents were established "in holiness and justice and in which man knew neither evil nor death," but that Adam's sin caused "human nature, common to all men, to fall into a state in which it bears the consequences of that offense, and which is not the state in which it was at first in our first parents . . ."

He explains the transmission of *Original Sin:*

> It is human nature so fallen, stripped of the grace that clothed it, injured in its own natural powers and subjected to the dominion of death, that is transmitted to all men, and it is in this sense that every man is born in sin. We therefore hold, with the Council of Trent, that original sin is transmitted with human nature, "not by imitation, but by propagation" and that it is thus "proper to everyone."

And ***Redemption:*** "We believe that Our Lord Jesus Christ, by the sacrifice of the Cross, redeemed us from original sin and all the personal sins committed by each one of us . . ."

The Fall: The *Catholic Catechism* by Fr. Hardon (pp. 100-101) states:

> Since the beginnings of Pelagianism and up to the most sophisticated theories of rationalism, the Church has never wavered in her essential doctrine about man's original condition as he left the creative hand of God, and of what happened when the first man disobeyed his Creator.

It explains that "Augustine's doctrine on original justice, the fall, and original sin was many times confirmed by successive Popes."

It refers to the Second Council of Orange and then says:

> A thousand years later, the Council of Trent returned to the same subject . . .[and] . . .the Church's doctrine at Trent becomes more sharply defined. Thus "the first man Adam immediately lost the justice and holiness in which he was constituted when he disobeyed the command of God in the Garden of Paradise."

The Catechism says that Trent wished:

> to carefully distinguish between two states of man's existence, before and after the fall. Before the fall, Adam enjoyed the gift of integrity, which meant absence of the conflict we now experience between our natural urges and the dictates of right reason. After the fall Adam lost this gift for himself and his posterity, since even those who have

been regenerated in baptism are plagued by an interior struggle with their unruly desires and fears.

So, too, Trent repeated in more explicit terms what earlier Councils had taught. Adam was to have remained immortal in body, but, when he sinned, he became subject to death. Trent confirmed St. Paul's doctrine that Adam's sin injured not only Adam himself but also his descendants. The consequences of Adam's sin were not only death of the body, but also the loss of grace—spiritual death—which passed from one man to all the human race.

What is Original Sin?:

> As Aquinas was later to explain, the essence of original sin is the deprivation of what God would have conferred on all Adam's descendants if the first man had not sinned. It is not some inherent evil in what God produces. (*Catholic Catechism* by Fr. Hardon, p. 105.)

Trust in Bible truth has been eroded lately. The point of entry of the erosion is *Genesis*, particularly regarding Adam and Eve. From there it has spread through the Bible.

We conclude with the warning of Pope Leo XIII, which should be heeded by today's teachers of young minds: ". . . for the young, if they lose their reverence for the Holy Scripture on one or more points, are easily led to give up believing in it altogether."

Appendix B

THE CHURCH AND FATHER TEILHARD DE CHARDIN

There is one writer who has exercised such an influence upon Liberal Catholic thought, particularly since Vatican II, that we must single him out for particular attention. This writer is Teilhard de Chardin. (Cf. Hugh J. O'Connell, C.SS.R., Ph.D. in *Keeping Your Balance in the Modern Church.*)

The Catholic Priests' Association of England, in their Newsletter of January-March, 1972, published an article titled *A Periscope on Teilhard de Chardin* by Mgr. John W. Flanagan. It summarized how Teilhard's case stands in the Church. The following is a condensation thereof.

Firstly, the sequence of official moves and edicts:
1926—Fr. de Chardin's Superiors in the Jesuit Order forbade him to teach any longer.
1927—The Holy See refused the "Imprimatur" for his book, *Le Milieu Divin.*
1933—Rome ordered him to give up his post in Paris.
1939—Rome banned his work, *L'Energie Humaine.*
1941—de Chardin submitted to Rome his most important work, *Le Phenomene Humaine.*
1947—Rome forbade him to write or teach on philosophical subjects.
1948—de Chardin was called to Rome by the Superior General of the Jesuits who hoped to get permission from

the Holy See for the publication of his most important work, *Le Phenomene Humaine (The Phenomenon of Man)*. Publication had been prohibited in 1944, and the prohibition was renewed in 1948. Fr. de Chardin was also forbidden to take a teaching post in the College de France.

1949—Permission to publish *Le Groupe Zoologique* was refused.

1955—de Chardin was forbidden to attend the International Congress of Palaeontology.

Fr. de Chardin died suddenly this year.

1957—In April, all Jesuit publications in Spain carried a notice from the Spanish Provincial of the Jesuits that de Chardin's works had been published in Spanish without previous ecclesiastical examination and in defiance of decrees of the Holy See.

1962—The Monitum (or warning): A Decree of the Holy Office under the authority of Pope John XXIII warned that Teilhard's works are replete with ambiguities, or rather with serious errors, which offend Catholic doctrine. It urged those in authority to effectively protect especially the minds of the young against the dangers of the works of Fr. Teilhard de Chardin and his followers.

1963—The Vicariate of Rome (a diocese ruled in the name of Pope Paul VI by his Cardinal Vicar), in a decree, required that Catholic bookshops in Rome should withdraw from circulation the works of de Chardin, and also those books which favor his erroneous doctrines.

Conclusion: Popes Pius XI, Pius XII, John XXIII and Paul VI endeavored to prevent the spread of the modernistic errors of Teilhard de Chardin.

Secondly, the *Periscope* article explained how Teilhard's works were published after his death by "progressive" elements within the Church, and some outside it, in disobedience to Teilhard's superiors and the Holy See.

Thirdly, the *Periscope* says that the Modernists in the Church set in motion a systematic campaign to spread Teilhard's doctrines into seminaries, schools, colleges, convents, etc. The campaign hinged around three points:

(1) That the decree of 1957 and the Monitum of 1962 have been misunderstood, and are now disregarded by the Holy See.
(2) That Pope Paul VI made a statement praising the works of Teilhard as "indispensable."
(3) That, while certain points of Teilhard's works may be contested, on the whole his works are perfectly reconcilable with the Church's teaching and give a new, deep and exciting "insight" into Catholic theology.

The Answer to (1):
(a) The edict and the Monitum are crystal clear and cannot be misunderstood.
(b) In reply to a formal query, the Sacred Congregation for the Doctrine of the Faith stated: "The judgment and dispositions made by the Congregation concerning the writings of Teilhard de Chardin have not been changed. Thus the Monitum of 30th June, 1962, continues in effect." (March 8, 1967).
(c) Further re-affirmations (October 20, 1967, March 23, 1970 and August 4, 1971) coming from Apostolic Delegations, but on the instructions of the Congregation for the Doctrine of the Faith, remove all possibility of doubt on this matter.

The Answer to (2):
To clear up this point, the question was put to the Con-

gregation for the Doctrine of the Faith. The following reply was given by the Congregation through the Apostolic Delegation to Washington (March 8, 1967): "I can authoritatively inform you that the Holy Father has never, in public or private, made this statement, that Father de Chardin is good and very necessary for our times."

The Answer to (3):

One of the greatest scholars on de Chardin, respected theologian Cardinal Journet, gave his verdict on the works of Teilhard as follows:

> De Chardin's works are disastrous . . . his synthesis is logical, and it must be accepted or rejected as a whole; but it contradicts Christianity . . . If one accepts de Chardin's explanation one must reject the Christian notion of Creation, Spirit, God, Evil, Original Sin, the Cross, the Resurrection, Divine Love, etc. (*Nova et Vetera,* October-December, 1962).

Appendix C

FERAL CHILDREN

Evolutionists, who propose that some primates walked upright because they were "nearly man" or "just man," are presuming that a degree of humanness makes an animal stand up. The feral children teach a different lesson. Amala and Kamala, with many generations of true humanness in them, but reared in the wolf den, did not stand up. They walked on hands and knees; they ran on all fours with alternate rising of rump and head.

When captured in 1920 and cared for by missionaries at the Orphanage of Midnapore (India), they behaved like frightened, caged animals. They ate on all fours, lapping liquids, grabbing food with their mouths. For some time after capture, they disturbed the orphanage at night with their eerie howling at 10 p.m., 1 a.m. and 3 a.m., the times when wolves howled to neighboring packs.

Kamala was about eight, and Amala was 1-1/2 years old when they were taken to the orphanage, but Amala died less than a year later. She showed promise of learning language and human ways with somewhat less difficulty than the older Kamala, which is a point of some significance.

After about two years, Kamala stood on her knees supported by a table and used both hands to bring a bowl of rice to her mouth. It was considerably later that she learned to stand on two feet. She never ran on two feet. Even after several years of human care, she reverted to quadrapedal running when speed was necessary, and by this method she ran so fast it was hard to overtake her.

After six years she was fitting in with orphanage life. She had increased her vocabulary to 45 words, 50% better than it was twelve months before. She understood verbal instructions. She could say small sentences like "Dolls inside box," involving a preposition, and suggesting a maturity in language better than a normal two-year-old.

A year later, she was responsible enough to take care of the orphanage babies for short periods. She had come to love the children and they all loved her. She was then about 15 years old. Apparently she continued to make progress in speech. Two years later she became ill. It is recorded that, although she became very weak and suffered long, she talked a great deal, and "with the full sense of the words used." She died in November, 1929, nine years after leaving the wolf den.

Kamala being fed by Mrs. Singh
(*Impression from a photograph in* Harper's Magazine)

This shows progress in Kamala's regeneration. When Kamala and Amala were rescued from the wolf's lair, their only vocabulary was a howling. At the orphanage they crouched in a corner of their room. Food had to be left on the floor, and they would take it in their mouths. Kamala would roll her eyes and growl if anyone approached when she was eating.

It took more than six months before they ventured to take a biscuit from the hand of Mrs. Singh, snatching it by mouth and darting back to their corner.

Some real progress had been attained by the time of Amala's death. The loss of Amala brought the first tear to Kamala's eyes. She would not eat, and reverted to the wild state. Special attention by Mrs. Singh won Kamala back. The illustration indicated her growing confidence in Mrs. Singh.

Further reading:

Wolf Children and Feral Man by Rev. J. A. L. Singh and Robert M. Zingg. Hamden, Connecticut: Archon Books, The Shoe String Press, 1966.

Biography of a Wolf Child. Article by Arnold Gasell, M.D., *Harper's* Magazine, January, 1941.

The Wild Boy of Aveyron by Harlan Lane, Harvard University Press, 1976.

Philosophy in a New Key by Susanne K. Langer (Chapter V: Language), Harvard University Press, 1951.

Appendix D

ADDENDUM TO "HUMAN EVOLUTION"— AND A WORD ON BIRDS

Chapter 4 of this book discussed, among other matters, two recent discoveries by the Leakey/Walker team which have shaken the evolution story.

Firstly, the July 1986 finding of a skull of A. Boisei (or Zinjanthropus). Its dating of 2-1/2 million years made it too old to fit the story. It wiped the slate of "hominid" evolution fairly clean.

Also, the finding in 1984 of the skeleton of "The Boy": The skull suggested Neanderthal (sapiens). The postcranial skeleton (which is the skeleton below the skull) essentially resembled modern man. However, the dating was 1.6 m.y. (million years) and this made "The Boy" too old to be classed Homo sapiens in evolution theory framework. The dilemma was skirted by classing him in the convenient group of Homo Erectus.

Now let us look at a third discovery, which was not discussed in Chapter 4—a recently reported discovery by a Johanson expedition.

In July, 1986, they discovered parts of the skull and some limb bones of an individual. The find is labeled OH62. It was reported in *Nature* (May 21, 1987, pp. 205 *ff.*). Teeth and skull parts indicated that OH62 is Homo Habilis.

Problems arose with the limb bones. *Nature* stated: "This represents the first time that limb elements have been securely assigned to Homo Habilis." (p. 209).

Homo Habilis had been regarded as an advanced hominid. But these bones show a very small creature barely three feet tall. This is as small as, or smaller than Lucy (A. Afrarensis). Its long arms, with fingers reaching almost to the knees, showed it to be as ape-like as Lucy. Yet this Homo Habilis was dated 1.8 m.y., which is 2 m.y. *after* Lucy. Here was evidence that, in the supposed 2 million years from Lucy to Habilis, nothing had happened. Evolution had stood still.

As we pause and contemplate the above three discoveries, we note firstly that the A. Boisei fossil wipes the slate clean of evolutionary ancestors.

Next, if we use Lucy as a datum point 3.8 m.y. ago, we go forward 2 million years to Homo Habilis (OH62) and we note that there is no evolutionary progress at all.

However, a brief 200,000 years later, in a miracle of evolutionary speed, The Boy suddenly appears, and he looks like Homo sapiens.

After The Boy, we go forward a long 1.6 million years to our own time, and we find our own postcranial skeletons are essentially the same as that of The Boy. This means that there was no real evolution in 1.6 m.y. from The Boy to us.

Therefore, as we look across nearly four million years from Lucy to us, there has been no evolution except for that one miraculous burst which produced The Boy. This wrecks the evolution story, and evolutionists have yet to explain it.

The facts and significance of the OH62 find are set out clearly in an article in *Impact*, No. 171 (September, 1987), by Duane T. Gish, Ph.D.

It is of further interest that the article also discusses the finding of two fossil *birds* dated 225 m.y. old, which was announced in *Nature* 322:677 (1986). On such dating, these very real birds would have lived 75 m.y. before Archaeopteryx. That should effectively refute any lingering claims about Archaeopteryx being a link between reptiles and birds.

Of course, we do not accept evolutionists' datings and the postulated great ages on which they have woven evolution's story, but it seems that every recent discovery has made their own evolution story more impossible.

Any Questions?

SELECTED QUESTIONS ANSWERED

Q. How could the Ark possibly have carried all the animals necessary?

A. This question is handled at length in *The Genesis Flood* by Morris and Whitcomb. If we assume 17-1/2 inches for a cubit, the Ark would have been 437 feet long by 73 feet wide and 44 feet high—build like an enormous barge and almost uncapsizable. Its gross tonnage would have been 14,000 tons. It was, by far, the biggest vessel ever built until very recent times. The three decks would give a carrying capacity equal to 522 standard American railroad cars.

The Genesis "kinds" would not include all species, and certainly not varieties of species. Thus, the animals on the

Ark would be restricted to types or kinds. The Ark would not have carried fish or any aquatic creatures. The conclusion is reached that, at the very outside, the Ark would need to carry not more than 35,000 individual vertebrate animals. Most animals are smaller than a sheep. The young of very large animals could have been carried instead of the fully grown. Even allowing the average to be the size of a sheep, it is estimated that the 35,000 could have been fitted into 146 railroad cars.

The Ark would have easily carried the animals on one deck, leaving one deck for the humans, and one deck for storage.

Q. How could Noah round up all those creatures?

A. He could not have done it. We have to acknowledge that God did the mustering. The Bible makes this clear. It says that Noah and his family went into the Ark, and that all the creatures "went in to Noah into the ark . . . And they that went in, went in male and female of all flesh, as God had commanded him: and the Lord shut him in on the outside." (*Genesis* 7:14-16).

If we wonder about kangaroos and polar bears and other far-flung animals making the journey to the Ark, we have to realize that the evidence shows the whole earth used to enjoy a fairly uniform and mild climate, with no extremes; therefore there were no specialized creatures adapted to extremes of heat or cold. There probably were no polar bears because there were no frigid zones for them. All the then existing species of animals could have lived in proximity to the Ark.

A number of competent scientists believe that the earth was probably surrounded by a transparent vapor canopy, high in the stratosphere (the waters above the firmament), and that the canopy caused a greenhouse effect on earth and gave a uniformly mild climate.

Q. How could the menagerie be managed and fed in the Ark for more than a year?

A. In the case of very large animals and carnivorous animals, the difficulty could have been avoided by having only young specimens aboard.

Alternatively, God may have used mechanisms like hibernation and estivation to quiet the creatures and make constant feeding unnecessary.

Morris and Whitcomb raise the interesting thought that hibernation, estivation and migration are the three methods of coping with inclement climactic conditions; but, if there existed a constantly mild climate, there would have been no reason for the existence of any of the three mechanisms before the Flood. They then suggest that it may have been on the eve of the Deluge that these abilities were first imparted to the animals. Certainly divine power could have kept the animals in a quiescent state in the Ark to minimize their feeding and other supervision.

The Bible does assure us that God was directing events. It tells us, "And God remembered Noah, and all the living creatures, and all the cattle which were with him in the ark . . ." (*Genesis* 8:1). The Bible is not suggesting that God absentmindedly forgot, and then suddenly remembered that Noah and the Ark were still out there in the flood. The Bible passage makes sense if it means: "And God *protected* Noah, and all the living creatures," etc. Apparently the Hebrew word "remember" can mean "protect."

Morris and Whitcomb tell us: "According to Hebrew usage, the primary meaning of 'Zakar' (remember) is 'granting requests, protecting, delivering' when God is the subject and persons are the object."

Q. Where would the water come from for a worldwide deluge?

A. Under our present conditions there is not enough water in the atmosphere to sustain 40 days and nights of global rain. In fact, if it were all precipitated, it would cover the ground to a depth of less than two inches.

There is compelling geologic evidence that a global flood did happen and that the highest mountains have been submerged. We cannot dodge the problem by saying that the flood never happened. Where, then, did the water come from?

The *vapor canopy* referred to in answer to the second question would be part of the solution. Another source would be *"juvenile waters,"* that is, waters which are added to the oceans through volcanoes, hot springs and other vents. Even today there is at least a cubic mile of such water added to the oceans each year. The Deluge was an unprecedented upheaval with volcanic action unimaginable. This would have added vast amounts of juvenile waters to the earth's surface.

Then, volcanic dust flung to the upper atmosphere could have provided nuclei of precipitation for the vapor canopy, whereupon its waters began raining on to the earth.

In the six hundredth year of the life of Noah, in the second month, in the seventeenth day of the month, all the fountains of the great deep were broken up (submarine volcanoes?) and the flood gates of heaven were opened (vapor canopy?) and the rain fell upon the earth 40 days and 40 nights. (*Genesis* 7:11-12).

Yet, even those sources would not suffice to cover mountains like Everest (29,000 ft.) or even Ararat (17,000 ft.). What we have to understand is that at the time of the Deluge there would not have been such high mountains for the Deluge to cover. Topography depends on the principle of "Isostasy" (equal weights). Somewhere, deep in the earth's crust, is a datum line; and, for equilibrium, the weights above the line have to balance. Areas of high topography must be of low density, and vice versa. Before the Deluge, the amount of water was much less than now; therefore the weight of oceans could balance only relatively low mountains. "Mountains were relatively low and ocean beds relatively shallow as compared with present conditions." (*Genesis Flood,* p. 268).

Even though the mountains were fairly low, yet more water was needed to submerge them, and from the oceans themselves came the greatest flooding. It is known that Europe was covered by the sea during man's history, and even the high plateau of Iran was devastated by sea water. All the continents bear evidence of having been submerged by sea water. The great coal deposits were laid down under sea water. Geologists would explain continental inundation as due to depression of the land, and there is good reason to couple this with an accompanying elevation of the bottom of the sea as it heaved to great volcanism and earthquakes.

In the Noahic cataclysm, water came down from the skies, came up from subterranean depths, and the oceans rose to engulf the land, while volcanoes and earthquakes caused colossal tidal waves which came and went around the drowned planet. Eventually, all this water had to be gotten off the land.

The Bible specifically refers to "the fountains of the great deep," so we infer that the greatest volcanic activity was sub-oceanic. The ejected lavas and juvenile waters would leave behind them great voids in the earth's crust, deep below the ocean beds. The weakened ocean beds could not support the vast increase in surface water and the great sediments washed down from the land. The ocean beds would sink under the burden; and correspondingly, the continental blocks would be forced upwards. This would have been the mechanism whereby the flood waters were removed from the land areas.

It is recognized by geologists that nearly all the great mountain areas of the world have Pliocene and Pleistocene fossils near their summits, which means that they were uplifted recently, and essentially simultaneously. (*Genesis Flood*, p. 128).

Geologists recognize that there have been "recent" rises of thousands of feet in mountain systems in Europe, America and Asia; and that high volcanic cones of the Pacific,

Asia and eastern Africa are believed to have been built up in the recent past. It is worth mentioning that Mt. Ararat's lava was deposited under water.

It should be explained that Creationists do not accept the terms Pliocene and Pleistocene in the "millions-of-years" context; but, as designations, they refer to recent times. (Refs. *The Genesis Flood; Scientific Creationism; Science of Today and the Problems of Genesis.*)

Q. How did the races of man originate?

A. For races to begin, evolutionists and creationists both agree that the prerequisite is inbreeding in a small, isolated group of people.

Dr. Morris, in *Scientific Creationism,* quotes Ralph Linton of Yale, a leading anthropologist and evolutionist, who explained in 1955:

> Observation of many different species has shown that the situation of small, highly inbred groups is ideal for the fixation of mutations and consequent speeding up of the evolutionary process. In general, the smaller the inbreeding group, the more significant any mutation becomes for the formation of a new variety.

Dr. Morris points out that mutations are harmful, not helpful, and would most likely destroy the population before effecting any imaginary benefits. However, if we change the word "mutations" to "recessive genes," creationists would then agree with Linton's statement.

In large populations, the population generally exhibits the characteristics of dominant genes. Only when a small group is isolated and interbreeds do the recessive genes have an opportunity to become typical.

Apparently there is no need for slowly developing racial distinctions over long periods of time. Rather, small inbreeding groups, exhibiting recessive gene characteristics, can effect distinct physical changes quite rapidly.

To produce the major racial divisions there is the question of what, in man's early history, caused mankind to disperse into small groups. The evolutionist cannot supply an answer, but creationists have an obvious explanation. Communication is a fundamental need in a group, and communication is by language. If a large group with a common language found that its language was suddenly fragmented into various languages, communication among the various sub-groups would become impossible. The large group would have to split into smaller groups according to language. Divisions of language would achieve the prerequisite of small, self-contained groups, whose inbreeding would produce the races.

Dr. Duane Gish has commented that when language was confused at the Tower of Babel, people would have dispersed in their lingual groups to different lands, probably in fairly small groups which would then inbreed in isolation. He suggests that God may have bestowed languages deliberately so as to marshal genetically similar individuals into the same language group. Thus, those individuals having a higher proportion of genes for Negroid features may have been given a common language, and similarly those who tended to Caucasian traits.

Q. Are we to believe that men lived for hundreds of years, as *Genesis* says?

A. Evidence shows there was a prehistoric period when

the whole earth had a temperate climate. Many believe that this was due to a vapor canopy above the stratosphere causing a greenhouse effect. Uniform temperateness would mean no strong wind currents, no storms. Plants and animals, including representatives of today's species, were giant-sized, and there is evidence of large stature for at least some of early mankind. It was a world vastly different from today's world. In that pre-Flood world the Bible records human lifespans of many hundreds of years.

In an article in *C.R.S. Quarterly* (June, 1978), Joseph C. Dillow says that a vapor canopy of magnitude sufficient to produce (during the Deluge) heavy rain for 40 days and nights would have caused a pre-Flood atmospheric pressure about double that of today, with about double today's oxygen pressure.

Higher oxygen pressure is beneficial to biological systems. In Florida, hyperbaric treatment using 2.5 atmospheres of pure oxygen has relieved effects of aging, helped treatment of strokes, improved memory and energy. Such pressurized pure oxygen is greater than the atmospheric oxygen pressure under the assumed pre-Flood canopy, but Dillow suggests that the latter, when extended over a whole lifetime, might have had similar beneficial effects in retarding senility.

Kevin C. McLeod, in *C.R.S. Quarterly* (March, 1981), points out that medical investigators have applied *electromagnetic fields* to a variety of patients with apparently beneficial effects including retarding of aging and stabilization of the genetic code, and also increased release of calcium into tissues. A relevant point is that disturbed calcium metabolism is a suspected factor in aging. With bone fractures that would not join, electromagnetic fields promoted bone growth and caused bone ends to unite and knit.

On the evidence, the earth's magnetic field is decaying exponentially. In the pre-Flood era it would have been very much stronger than now. People in that era would have enjoyed the benefits of a much greater electromagnetic field, presumably with effects on longevity.

Donald W. Patten, in *C.R.S. Quarterly* (June, 1982), looks at the role of *carbon dioxide*. In laboratory experiments, an atmosphere enriched in CO_2 produced beneficial effects on the blood of vertebrate animals. Also, it caused dilation of blood vessels in the brain (and skin), making more oxygen available to brain cells. There is a small gland in the brain called the hypothalamus, a gland which affects aging for the neuro-endocrine system. Increased oxygenation in brain cells reduces the activity of this gland and thus reduces its influence for aging.

Patten proposes that the pre-Flood atmosphere was very much richer in CO_2 than was the atmosphere after the Flood. Why? Because cold oceans soak up much more CO_2 from the atmosphere than do warm oceans. Today's oceans average a chilly 38°F, compared with warmer pre-Flood oceans of perhaps 60°. The warmer oceans meant the pre-Flood atmosphere was much richer in CO_2, which would have resulted in dilation of the blood vessels, increasing oxygen flow, and thus would have rendered the hypothalamus less active and thereby retarded the aging process.

In an interesting aside Patten says that, a century ago, CO_2 comprised 290 parts per million of the atmosphere. Since then, increasing burning of fossil fuels has raised the CO_2 ratio to 330 p.p.m. He thinks this increase in atmospheric CO_2 has some relation to recent generations' increase in height and/or lifespan.

Fossils show that, before the Pleistocene Age, the size of mammals was 30% to 40% greater than in today's world. This giganticism occurred worldwide. Then, with the Pleistocene, which we interpret as the post-Flood world, there occurred a declining size of animals in all parts of the world. The fossils cannot reveal whether there was also a decline in lifespans of animals, but *Genesis* records a decline in man's lifespan.

Both Dillow and Patten draw attention to the fairly constant lifespans of the long-lived pre-Flood patriarchs from Adam to Noah, and then to the declining ages of men after the Flood. From Noah's son, Shem (600 years), through 17 generations to the contemporaries of Moses when 70 years became the ordinary lifetime, the lifespans plotted graphically against the generations show an exponential decline. Dillow comments that such a decay curve is common when a system in equilibrium is suddenly acted on in a way that shifts it to a new equilibrium. He says that *myths* could not produce such a neat mathematical result. It is most unlikely that such a curve could result from anything but an actual *historical* happening. The decay curve "suggests that new factors were present in the post-Flood environment."

Oxygen, carbon dioxide, earth's magnetic field may all have played a part in longevity and in the mystery of aging. It is all in the investigatory stage, but these factors should persuade skeptics to think hard before dismissing the *Genesis* ages as myths.

Q. Who was Cain's wife?

A. This question is often asked, and sometimes in a tone that implies "Got ya' this time."

The answer is simple: Cain's wife was his sister. Then comes the objection that the Bible makes no mention of other children of Adam and Eve at the time Cain killed Abel. The Bible names Cain and Abel because it recounts an event concerning them. Its silence regarding additional children cannot be interpreted to mean that there were not other children.

The Douay version of the Bible is unquestionably Catholic. In a footnote explaining *Genesis* 4:14, the Douay Bible says regarding Cain:

> His guilty conscience made him fear his own
> brothers and nephews; of whom, by this time, there

might be a good number upon the earth; which had now endured near one hundred and thirty years; as may be gathered from Genesis 5:3, compared with Genesis 4:25, though in the compendious account given in the Scriptures, only Cain and Abel are mentioned.

Another footnote in the Douay Bible explains *Genesis* 4:17 which refers to Cain's wife. The footnote says: "She was a daughter of Adam, and Cain's own sister; God dispensing with such marriages in the beginning of the world, as mankind could not otherwise be propagated."

This usually provokes a further objection that God would not permit incest. However, the Bible clearly tells us that God started the human race with one couple, Adam and Eve. Unless God intended the human race to stop after one generation, God intended brothers and sisters to marry at this stage.* Before we express disappointment with God for allowing this, let us look at [one reason] why we regard incest as reprehensible.

We humans carry what is called "the genetic load." This is the accumulation of bad mutations during the centuries. Fortunately for us, the genetic effect of these mutations is usually recessive. It remains latent, unless both parents carry the particular recessive gene. In that case the offspring will probably exhibit the defect. If parents are closely related there is greater risk that both will carry a matching reces-

---

*For clarification, we refer the reader to the remarks by Fr. Austin Fagothey, S.J. in *Right and Reason*, 2nd. ed. (St. Louis: C. V. Mosby, 1959; TAN, 2000, p. 375-6). Fr. Fagothey states that whereas marriage between parent and child is absolutely against the nature law, marriage between brother and sister is not *absolutely* contrary to the natural law, but is under extremely stringent conditions. He states that "only God could allow it, and He would do so only if otherwise the race could not propagate." Fr. Fagothey sums up the reason for the wrongness of brother-sister marriage by stating that it would mean "the utter ruin of the family and make the home an unlivable place." —*Publisher*, 2000.

sive gene from the genetic load; and so, the risk of defective children is greater.

Incest increases the genetic risk, but does not necessarily mean defective children. Ancient Egyptian ruling families practiced brother-sister marriages and produced healthy kings and queens. This is mentioned by Ashley Montagu, author of *Human Heredity;* and he gives other examples, such as the inhabitants of the Pitcairn Islands, the Hindu community of Tengger Hills and people of many small islands. All these seem to show no ill effects. On the other hand, inbreeding among the Nanticoke Indians of Delaware produced a drooping upper eyelid; and inbreeding in the population of Martha's Vineyard was the cause of deafness in the hill folk of New England and of considerable feeblemindedness. (Ref. *Supplement to Bible Science Newsletter,* April, 1975).

Now we come to the main point of our answer. Adam and Eve were bodily perfect. In the early stages of the human race there was virtually *no genetic load.* When Cain took his own sister as wife, both were children of Adam and Eve. There was no genetic risk to their children.

Philosophically, let us add that God's plan was wise. He started humanity with one couple; thus the whole human race are brothers and sisters. In starting us the way He did, God was fully aware that there would be no genetic risk from marriages of close relatives among the early generations.

Q. In a high school class, a leaflet was distributed saying that new research on chromosomes shows that humans and chimpanzees differ surprisingly little; that the great apes have 48 chromosomes and humans have 46, that essentially every band and sub-band observed so far in man has a direct counterpart in the chimp chromosomes. The leaflet says that our common ancestor probably also had 48, but, during our evolution, two of

these fused to form what is now chromosome No. 2 in humans.

The question is: Is this new evidence of evolution of man?

A. The leaflet states some facts which are correct, but it adds assumptions which are only suppositions, e.g., the assumption that evolution is fact and the assumption of some hypothetical, unidentifiable "common ancestor."

We have to keep in mind that man has 46 chromosomes in 23 pairs, the chimpanzee has 48 chromosomes in 24 pairs.

Regarding chromosomes of chimps and man, the late Professor Jerome Lejeune, of Paris University, was a world authority. Professor Lejeune stated that chromosomal research clearly demonstrates that the genetic differences between man and each of the three great apes are so great as to provide conclusive evidence that man did not evolve from his closest kin, the apes. There are as many chromosomal differences between man and each of the apes as there are between any one ape species and another. In Australia in 1978 Professor Lejeune stated:

> We now know, thanks to the work of one of my assistants, that the chimp has two chromosomes more than we have. The chimp has two chromosomes which are separated. Man has a big chromosome which is made by the joining of the analogous two chromosomes of the chimp.

My interpretation is that, where Professor Lejeune mentions two chromosomes of the chimp, he is referring to two *pairs*. Then two *pairs* of ordinary chromosomes in the chimp have the equivalent of *one big pair* of chromosomes in man.

He explained that the joining of the two chromosomes is head to head, which, until recently, had been regarded as impossible. When they are thus joined, the genetic information of the second chromosome in the chimp is read in one direction, but its fused counterpart in man is read in the reverse direction. The reading of the information in the chimp's direction may give one sense, but, when read in the human way, it gives a different significance.

If a gene contains 1,000 or more nucleotides, and if a nucleotide directs the position of an amino acid, and if one amino acid out of position can cause biological havoc, let us imagine the effect of the reversal of a chromosome containing thousands of genes. When such immensity of genetic information can be read forwards (for a chimp) and backwards (for man) without biologically wrecking the chimp or the man, it suggests clever design by a super-intelligence.

Professor Lejeune affirmed that research since 1971 has shown that the Darwinist idea of evolution by gradual change is genetically impossible. He is definite that the only way anything could have evolved is by sudden and complete breaks. That means evolution by big jumps, so we are looking at the "hopeful monster" idea again.

Having established that man, chimpanzee, gorilla and orangutan are equally far apart, and none of them could have evolved into another, Lejeune concludes thus:

> A simplified theory might suppose that all four came from a common ancestor, through different species that were separated long ago, and that the common ancestor was not an ape at all, but some small mammal.

The scientific position is clear: Science observes man and three species of ape, and science pronounces that man could not have evolved from any ape. That is all that science can tell us. Scientists can hypothesize all sorts of things if they desire evolution. So some scientists (and some teachers) are hypothesizing that evolution of man did happen and that man and chimp have evolved along separate lines from an unknown "common ancestor."

In body structure there is some rough similarity between man and chimp, so it is not surprising that there is a considerable similarity in chromosomes. However, even if the only difference were in that fused chromosome in man, that would involve some thousands of genes of human genetic information as opposed to chimpanzee information; and that constitutes a world of difference. Lejeune reminds us that our bodies are human because the genetic information that molded our bodily material is human information. "Otherwise," he says, "we would be flies or chimpanzees."

If you want to believe in evolution, you have to abandon evolution by gradual steps. You must believe in sudden and complete breaks. You have to accept evolution by "monsters" which (instead of dying as all monsters do) survive and launch new species; and you must believe that these "hopeful monsters" have been happening so frequently as to produce the innumerable species that have ever lived on earth.

So frequent a happening could not stop now. Your pet mare's expected foal might be something not a foal, but a something never before seen on earth.

To be consistent, you must not be surprised if, someday, your own child is not the expected baby but something other than human, never before seen on earth, and that this little monster will survive, but be unable to breed with humans. Lejeune has said that, to start a new species, there have to be at least two of these. Before your own monster can breed a new species, a second monster has to be born about the same time, one of opposite gender, with complementary reproductive organs.

Evolutionists like to hypothesize back into the dim, untestable past. If you play that game, you must ask yourself: Might it not happen, just as easily, in my own suburb, in my own home, at any time?

I know, and you know, that it will not happen.

SOME FURTHER RESOURCES
Updated by the Publisher, 2000.

Catholic Resources

The best and most recent Catholic treatment of evolution is *Creation Rediscovered: Evolution and the Importance of the Origins Debate*, by Gerard J. Keane (Rockford, IL: TAN, 1999). This book exposes the problems with evolution vis a vis both science and the Catholic Faith, and it includes an extensive bibliography.

An audio tape by Wallace Johnson, plus the following other tapes on evolution by Catholic authors, are available from Keep the Faith, P.O. Box 277, Ramsey, NJ 07446. Tel. 201-327-5900:

Evolution: The Hoax that's Destroying Christendom. Audio. Wallace Johnson.

Evolution: Fact or Belief? Video. Peter Wilder.

Evolution. Dr. William Marra. Audio.

Evolution: A Fairy Tale for Adults. Audio. Narrated by Fr. Valentine Long, O.F.M.

Watchmaker newsletter—"Official Magazine of Morning Star Catholic Origins Society" (At present this newsletter is not produced on a regular basis but is published occasionally):

Watchmaker
c/o Fr. David Becker
St. Stephen's
303 Lincoln Way E.
McConnellsburg, PA 17233
dbecker@innernet.net

Daylight newsletter
Daylight Origins Society
Anthony Nevard
19 Francis Ave.
St. Albans, Herts
England AL3 6BL

The following books of Father Patrick O'Connell are among the few dealing with the subject of evolution from a Catholic standpoint. Some of his materials would have to be modified in view of more recent research, but they are still valuable. Also, they give a good summary of Church documents and pronouncements on the subject.

Science of Today and the Problems of Genesis, TAN Books and Publishers, Inc., Rockford, IL.

The Origin and Early History of Man, Lumen Christi Press, Houston, Texas.

Original Sin in the Light of Modern Science, Lumen Christi Press.

Other Resources

Doorway Papers (No. 32), *Primitive Cultures, Their Historical Origins,* Part 1 and Part 2, by Dr. Arthur C. Custance. Brief and information-packed.

The Doorway Papers series collected in several volumes by Zondervan, Grand Rapids.

NOTE: The following were formerly available from Creation Life Publishers, San Diego, California. We understand that Creation Life Publishers are no longer in active operation but that a number of the following titles—those marked with an asterisk—are available from Institute for Creation Research (a Protestant ministry), P.O. Box 2667, El Cajon, CA 92021. Tel. 619-448-0900 and 800-999-3777. Other of these titles are apparently out of print. Another source for creation science materials is Answers in Genesis (a Protestant ministry and formerly known as Creation Science Foundation), P.O. Box 6330, Florence, KY 41022. Tel. 606-727-2222. Another possible source: Master Books, P.O. Box 727, Green Forest, AR 72638-0727. Tel. 800-643-9535.

The Origin of Life: A Critique, by Duane T. Gish.
 Critique of Radiometric Dating, by Harold S. Slusher, M.S.
 Origin and Destiny of Earth's Magnetic Field, by Thomas G. Barnes, Professor of Physics at University of Texas, El Paso.
Acts/Facts/Impacts, Volumes, being compilations of the "Impact Series" papers.
The Genesis Flood, 1961; 15th printing 1972, by Henry M. Morris, Ph.D. and John C. Whitcomb, Jr., Th.D.
The Twilight of Evolution, 1963, by Dr. H. M. Morris.
Scientific Creationism, 1974, by Dr. H. M. Morris.
 Biology: A Search for Order in Complexity, 1973, by scientists of Creation Research Society. Biology textbook for high schools.
 The Origin of the Solar System, 1964, by John C. Whitcomb, Jr., Th.D.
 Why Not Creation? 1971; selected articles from *C.R.S. Quarterly.*
 Scientific Studies in Special Creation, 1971; more articles from C.R.S. Quarterly.
 Speak to the Earth, 1975; selected articles from *C.R.S. Quarterly,* 1969 to 1974.
 Darwin: Before and After, 1967, Robert E. D. Clark, Ph.D.
 Prehistory and Earth Models, 1966, Melvin A. Cook, Ph.D.
Evolution: The Fossils Still Say No! Duane T. Gish, Ph.D. (Earlier editions: *Evolution: The Fossils Say No!* [1973] and *Evolution: The Challenge of the Fossil Record* [1985]).

Evolution and Christian Faith, 1969, Bolton Davidheiser, Ph.D.

The Transformist Illusion, 1955, Douglas Dewar, F.Z.S.

**It's A Young World After All,* 1986, Paul D. Ackerman, Ph.D.

**Darwin's Enigma,* 1984, by Luther D. Sunderland, B.S.

The Creation Explanation, 1975, Robert E. Kofahl, Ph.D., and Kelly L. Segraves, Hon. Doctorate in Science.

In Australia, the above may be obtainable through Creation Science Foundation, P.O. Box 302, Sunnybank, Queensland, 4109. (New name and Australian address: Answers in Genesis, Acacia Ridge D.C., 4110 Qld., Australia.)

Also:

Man: Ape or Image, 1981, by Dr. John Rendle-Short.

**Bone of Contention,* by Silvia Baker, M.Sc.

Ape-Men: Fact or Fallacy? 1977. A penetrating examination by Malcolm Bowden.

In England, The Creation Science Movement (formerly called The Evolution Protest Movement) is a source of good literature exposing the non-science of evolution. It has been producing solidly scientific anti-evolution literature for half a century. The Creation Science Movement's address is: The Secretary, Creation Science Movement, 20 Foxley Lane, High Salvington, Worthing, W. Sussex BN13 3AB, England. (See p. 204 for new or alternate address.) Two very readable publications of The Creation Science Movement are:

A Challenging "Introduction" to The Origin of Species, 1956, by Professor W. R. Thompson, F.R.S.

Lessons of Piltdown (A Study in Scientific Enthusiasm at Piltdown, Java and Peking), 1959, by Francis Vere.

In the U.S.A., Creation Moments (formerly The Bible-Science Association, Inc.) handles a large selection of literature and videos exposing the weaknesses of evolution. Address: P.O. Box 260, Zimmerman, MN 55398-0260. Tel. 800-422-4253.

Periodicals:

Creation Research Society Quarterly, P.O. Box 8263, St. Joseph, MO 64508-8263. Tel. 816-279-2626.

Creation Moments (formerly titled *Bible-Science Newsletter*), P.O. Box 260, Zimmerman, MN 55398-0260.

*I.C.R. Acts and Facts and its supplement *Impact* by The Institute for Creation Research, P.O. Box 2667, El Cajon, CA 92021.

Creation Ex Nihilo, Quarterly of The Creation Science Foundation, P.O. Box 302, Sunnybank, Queensland 4109, Australia. (New name and U.S. address: Answers in Genesis, P.O. Box 6330, Florence, KY 41022.)

Creation Science Movement Bulletin, 20 Foxley Lane, High Salvington, Worthing, W. Sussex BN13 3AB, England. (New or alternate address: 50 Brecon Ave., Portsmouth, U.K. PO6 2AW).

Non-Creationist works referred to in this book:

Implications of Evolution, Dr. G. A. Kerkut, London. Pergamon, 1960.

Readings in Earth Sciences, Editor: Dr. Ben Moulton. Von Nostrand Reinhold Co., 1972.

Beyond the Ivory Tower, Solly Lord Zuckerman. Taplinger Publishing Co. of New York, 1970.

The Neck of the Giraffe, Francis Hitching (Royal Archaeological Institute member; Prehistoric Society member). Pan Books Ltd., London, England, 1982.

Also recommended:

*Evolution: A Theory in Crisis, by Dr. Michael Denton. Burnett Books, London, 1985; Adler & Adler, Bethesda, MD, 1986. U.S. distributor: Woodbine House, Bethesda, MD.

If you have enjoyed this book, consider making your next selection from among the following . . .

Prices subject to change.

Angels and Devils. *Joan Carroll Cruz* 15.00
Moments Divine—Before the Blessed Sacrament. *Reuter* 8.50
Miraculous Images of Our Lady. *Cruz* 20.00
Miraculous Images of Our Lord. *Cruz* 13.50
Raised from the Dead. *Fr. Hebert* 16.50
Love and Service of God, Infinite Love. *Mother Louise Margaret* 12.50
Life and Work of Mother Louise Margaret. *Fr. O'Connell* 12.50
Autobiography of St. Margaret Mary. 6.00
Thoughts and Sayings of St. Margaret Mary 5.00
The Voice of the Saints. *Comp. by Francis Johnston* 7.00
The 12 Steps to Holiness and Salvation. *St. Alphonsus* 7.50
The Rosary and the Crisis of Faith. *Cirrincione & Nelson* 2.00
Sin and Its Consequences. *Cardinal Manning* 7.00
St. Francis of Paola. *Simi & Segreti*. 8.00
Dialogue of St. Catherine of Siena. *Transl. Algar Thorold* 10.00
Catholic Answer to Jehovah's Witnesses. *D'Angelo*. 12.00
Twelve Promises of the Sacred Heart. (100 cards). 5.00
Life of St. Aloysius Gonzaga. *Fr. Meschler* 12.00
The Love of Mary. *D. Roberto* 8.00
Begone Satan. *Fr. Vogl* 3.00
The Prophets and Our Times. *Fr. R. G. Culleton* 13.50
St. Therese, The Little Flower. *John Beevers* 6.00
St. Joseph of Copertino. *Fr. Angelo Pastrovicchi* 6.00
Mary, The Second Eve. *Cardinal Newman* 3.00
Devotion to Infant Jesus of Prague. *Booklet* 75
Reign of Christ the King in Public & Private Life. *Davies*. 1.25
The Wonder of Guadalupe. *Francis Johnston* 7.50
Apologetics. *Msgr. Paul Glenn* 10.00
Baltimore Catechism No. 1. . . . . 3.50
Baltimore Catechism No. 2. . . . . 4.50
Baltimore Catechism No. 3. . . . . 8.00
An Explanation of the Baltimore Catechism. *Fr. Kinkead*. 16.50
Bethlehem. *Fr. Faber* 18.00
Bible History. *Schuster* 13.50
Blessed Eucharist. *Fr. Mueller* 9.00
Catholic Catechism. *Fr. Faerber* 7.00
The Devil. *Fr. Delaporte* 6.00
Dogmatic Theology for the Laity. *Fr. Premm*. 20.00
Evidence of Satan in the Modern World. *Cristiani* 10.00
Fifteen Promises of Mary. (100 cards) 5.00
Life of Anne Catherine Emmerich. 2 vols. *Schmoeger*. 37.50
Life of the Blessed Virgin Mary. *Emmerich*. 16.50
Manual of Practical Devotion to St. Joseph. *Patrignani* 15.00
Prayer to St. Michael. (100 leaflets). 5.00
Prayerbook of Favorite Litanies. *Fr. Hebert* 10.00
Preparation for Death. (Abridged). *St. Alphonsus* 8.00
Purgatory Explained. *Schouppe* 13.50
Purgatory Explained. (pocket, unabr.). *Schouppe*. 9.00
Fundamentals of Catholic Dogma. *Ludwig Ott* 21.00
Trustful Surrender to Divine Providence. *Bl. Claude* 5.00
Wife, Mother and Mystic. *Bessieres*. 8.00
The Agony of Jesus. *Padre Pio*. 2.00

Prices subject to change.

Freemasonry: Mankind's Hidden Enemy. *Bro. C. Madden* 5.00
Fourteen Holy Helpers. *Hammer* 5.00
All About the Angels. *Fr. Paul O'Sullivan.* 6.00
AA-1025: Memoirs of an Anti-Apostle. *Marie Carré.* 6.00
All for Jesus. *Fr. Frederick Faber.* 15.00
Growth in Holiness. *Fr. Frederick Faber.* 16.50
Behind the Lodge Door. *Paul Fisher.* 18.00
Chief Truths of the Faith. (Book I). *Fr. John Laux* 10.00
Mass and the Sacraments. (Book II). *Fr. John Laux* 10.00
Catholic Morality. (Book III). *Fr. John Laux..* 10.00
Catholic Apologetics. (Book IV). *Fr. John Laux* 10.00
Introduction to the Bible. *Fr. John Laux* 16.50
Church History. *Fr. John Laux* 24.00
Devotion for the Dying. *Mother Mary Potter.* 9.00
Devotion to the Sacred Heart. *Fr. Jean Croiset* 15.00
An Easy Way to Become a Saint. *Fr. Paul O'Sullivan* 5.00
The Golden Arrow. *Sr. Mary of St. Peter.* 12.50
The Holy Man of Tours. *Dorothy Scallan* 10.00
Hell—Plus How to Avoid Hell. *Fr. Schouppe/Nelson.* 10.00
History of Protestant Ref. in England & Ireland. *Cobbett.* 18.00
Holy Will of God. *Fr. Leo Pyzalski* 6.00
How Christ Changed the World. *Msgr. Luigi Civardi.* 8.00
How to Be Happy, How to Be Holy. *Fr. Paul O'Sullivan* 8.00
Imitation of Christ. *Thomas à Kempis. (Challoner transl.)* 10.00
Life & Message of Sr. Mary of the Trinity. *Rev. Dubois* 10.00
Life Everlasting. *Fr. Garrigou-Lagrange, O.P.* 13.50
Life of Mary as Seen by the Mystics. *Compiled by Raphael Brown.* . . . 12.50
Life of St. Dominic. *Mother Augusta Drane* 12.00
Life of St. Francis of Assisi. *St. Bonaventure* 10.00
Life of St. Ignatius Loyola. *Fr. Genelli.* 16.50
Life of St. Margaret Mary Alacoque. *Rt. Rev. Emile Bougaud.* 13.50
Mexican Martyrdom. *Fr. Wilfrid Parsons.* 10.00
Children of Fatima. *Windeatt. (Age 10 & up)* 8.00
Cure of Ars. *Windeatt. (Age 10 & up)* 12.00
The Little Flower. *Windeatt. (Age 10 & up)* 8.00
Patron of First Communicants. (Bl. Imelda). *Windeatt. (Age 10 & up)* . . 6.00
Miraculous Medal. *Windeatt. (Age 10 & up)* 7.00
St. Louis De Montfort. *Windeatt. (Age 10 & up)* 12.00
St. Thomas Aquinas. *Windeatt. (Age 10 & up).* 6.00
St. Catherine of Siena. *Windeatt. (Age 10 & up)* 5.00
St. Rose of Lima. *Windeatt. (Age 10 & up)* 8.00
St. Hyacinth of Poland. *Windeatt. (Age 10 & up).* 11.00
St. Martin de Porres. *Windeatt. (Age 10 & up)* 7.00
Pauline Jaricot. *Windeatt. (Age 10 & up)* 13.00
Douay-Rheims New Testament. *Paperbound* 15.00
Prayers and Heavenly Promises. *Compiled by Joan Carroll Cruz.* 5.00
Preparation for Death. (Unabr., pocket). *St. Alphonsus.* 10.00
Rebuilding a Lost Faith. *John Stoddard* 15.00
The Spiritual Combat. *Dom Lorenzo Scupoli* 9.00
Retreat Companion for Priests. *Fr. Francis Havey.* 7.50
Spiritual Doctrine of St. Cath. of Genoa. *Marabotto/St. Catherine* 12.50
The Soul of the Apostolate. *Dom Chautard* 10.00

Prices subject to change.

At your Bookdealer or direct from the Publisher.
Call Toll-Free 1-800-437-5876.

Prices subject to change.

About the Author

The late J. Wallace G. Johnson (1916-1989) was an Australian Catholic layman and businessman. He made a special study of the subject of evolution versus creation for sixteen years. He lectured extensively in high schools (Catholic and state) and to mixed audiences. On occasion he addressed priests and student priests at Pius XII Seminary, Brisbane; Christian Brothers and student Brothers at Xavier Training College, Brisbane; and audiences of students at the University of Queensland, Brisbane.

In later years, for health and other reasons, Mr. Johnson did little active lecturing, but he continued to reach audiences worldwide through tape recordings and writings. He also did research and writing on other topics, such as the Holy Shroud of Turin and Our Lady of Fatima.

Wallace Johnson died of cancer of the pancreas on June 29, 1989, being survived by his wife, their two sons and by several grandchildren. A tribute to him distributed at the time of his death stated:

"It is a singular tribute to Wallace that this book popularized an unusually complex subject, written especially for other Catholic laymen and carefully documenting not only the bankruptcy of Evolutionism in scientific terms, but also its incompatibility with Catholicism. . . . Indeed, his exposé of Evolutionism for his fellow Catholics is a one-of-a-kind eyeopener on a subject disastrously misrepresented by evolutionists. . . . His work and memory continue to serve as inspiration to others. The potential for on-going fruits of his labors is inestimable. We thank God for having blessed our very troubled world with so stalwart a Christian soldier."